The Herring People

Other Books by Scott Renyard

Illustrated Screenplays

Who Killed Miracle? (2022)

The Pristine Coast (2023)

The Unofficial Trial of Alexandra Morton (2023)

Trial of an Iconic Species (2023)

Screenplay Collections

Pressure Point: A Series of Mishandled Events (forthcoming 2024)

Children's Books

The Flag That Flew Up (2021)

The
Herring People

an illustrated screenplay

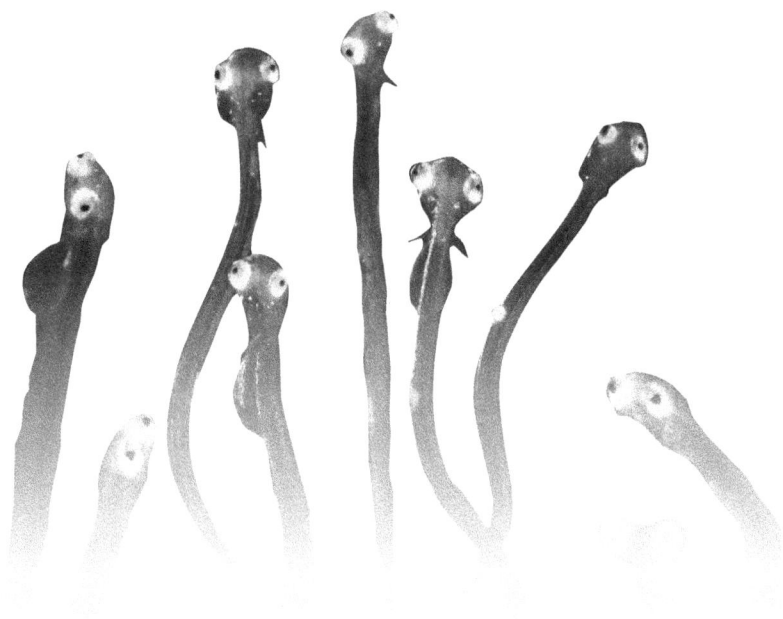

Scott Renyard

juggernaut CLASSICS

Published by Juggernaut Classics Inc.

Contact: scott@juggernautpictures.ca

Documentary copyright: 2023 Pacific Coast Entertainment Ltd.

All Rights Reserved.

ISBN: 978-1-998836-54-3 (softcover)
ISBN: 978-1-998836-55-0 (eBook)

Cover photography by Scott Renyard and Fernando Lessa
Edited by Lesley Cameron
Cover design by Caid Dow and Jan Westendorp
Book design by Jan Westendorp/katodesignandphoto.ca
"I Am the Herring" lyrics courtesy John McLachlan

Juggernaut Classics Inc.

Contents

Introduction

The Squamish Streamkeepers was formed over 20 years ago by a group of citizens who came from a wide range of backgrounds and shared a concern about the declines in wild salmon populations that are native to the Squamish River watershed. They met regularly to discuss the problem and to plan activities to help address it—and to ruminate over why their activities weren't as successful as they had hoped or expected. Over the years, some of the old-timers in the group began to talk about the dramatic decline in the local herring population that had happened back in the 1970s. Herring used to spawn in great numbers throughout the Squamish River estuary, and some of the streamkeepers remembered when the whole community would catch them right off the dock. As they talked, they realized that this decline corresponded with the decline in the local salmon populations. Given that herring is a major food source for young salmon, and the projects to help improve salmon habitat didn't seem to be making much of a difference, the streamkeepers began to wonder if they should expand their work to include Pacific herring. They then embarked on an initiative, led by Jonn Matsen, who later became known as the Herring Coordinator, to investigate whether the disappearance of herring from the Squamish Estuary was linked to the decline of the salmon.

The streamkeepers hoped that if they could bring back herring, they could also help the local salmon populations. Their first step was to look at ways to encourage the herring to spawn. To do this, they drew on the knowledge of local First Nations.

Traditionally, First Nations fishers would watch the waterways for signs of herring spawning activity and then place hemlock boughs or small saplings in the water. The herring would spawn on the hemlock and the fishers would remove the branches once the spawn was over and then harvest the eggs. The streamkeepers were concerned about the degraded condition of the Mamquam Blind Channel due to decades of industrial activity, right where some of them used to catch herring as kids. They decided to adapt the technique of the First Nations by submerging small hemlock trees not as a way to harvest the eggs but as a place to deposit them. Unfortunately, the experiment didn't work because the streamkeepers had overlooked the fact that herring would need to be actively spawning in the area for them to make use of the hemlock. But as luck would have it, a few weeks later, Matsen got a call from a guard at the Squamish Terminals, a shipping terminal located in the middle of the Squamish Estuary. The guard told him that he could see herring spawning along the shoreline near the guard shack. By the time the streamkeepers arrived, though, the herring were gone. But they had left eggs behind on the rocks, rockweed, and sea grasses— and suddenly the herring enhancement project seemed like it had a decent chance of success.

The streamkeepers then discovered salmon-rearing pens on the north end of the Squamish Terminals east dock. These pens were being used by the Department of Fisheries and Oceans (DFO) as part of an ongoing DFO project to try to improve the survival of salmon that were being raised at the Tenderfoot Creek hatchery, located on the Cheakamus River within the Squamish River watershed. The Tenderfoot hatchery, built in 1981, was one of many hatcheries started in BC to try to rebuild wild salmon populations. In this case, there was added concern about Squamish watershed chinook salmon, which had declined from an average of 35,000 spawners in the 1960s to a meagre 1,500 by the 1980s. The streamkeepers asked the Terminals management and DFO if they could move their hemlock experiment to the salmon pens to see if the herring would spawn on them. Permission was duly granted, and half a dozen small hemlock trees were lowered off the rearing pen wharf and into the water.

A couple of weeks later, the streamkeepers returned to the salmon pens at low tide and pulled up the hemlocks. There were no herring eggs attached to them, but a dozen or so small eels flopped off the branches and onto the platform. It looked like the eels had been using the hemlock branches as protective cover. (I have often wondered if this was a reflection of the lack of habitat in the estuary.)

As they stood on the salmon pen wharf, the streamkeepers noticed that the creosote pilings under the Terminals east dock were not black, as creosote pilings usually are. Instead, they looked as if they were covered in yellow slime. Had the streamkeepers not checked on their experiment at low tide, they might not have seen the slime on the pilings. But they did, and they were invested enough to go under the dock to investigate. There they discovered that the yellow slime was actually millions of rotting herring eggs. This discovery changed the direction of the streamkeepers' project and made me realize the story of their experiments could make an interesting film. At the time, I couldn't help but think about the fact that the Squamish Terminals were built right around the time the herring disappeared in the Squamish estuary. Was this coincidence or was there another factor behind the disappearance of herring in this location in the 1970s?

The streamkeepers assumed that the herring eggs had died from exposure to the creosote-soaked pilings. Creosote has long been used to protect pilings from various wood-boring organisms, the most destructive of which is a mollusk called the shipworm. These organisms are sometimes referred to as the termites of the sea and are so destructive that wood structures infested with them have been known to collapse. It appeared to the streamkeepers that while the creosote was protecting the pilings, it was also having a very detrimental effect on the estuary's herring population.

The streamkeepers publicized their discoveries through local media outlets and could have left the problem for the authorities to address. But they had a reputation as a group for taking action, and Matsen decided they should try to do something about the problem. He came up with the idea of wrapping the pilings with something that wasn't toxic to herring eggs. So, prior to the 2008 spawning season, he went to a local building supply store and bought rolls of black plastic, screen door mesh, and weed control cloth.

The streamkeepers got permission from the Squamish Terminals management to wrap pilings under the east dock with each of the three materials to see if any of them would protect herring eggs from the creosote. A couple of weeks later, they returned to find that the herring had found the wrapped pilings and spawned heavily on the weed control cloth. The streamkeepers knew that herring eggs only take 10–21 days to develop and hatch in nature. So they closely monitored the eggs on their wraps and were very excited when they all appeared to have hatched out.

This success encouraged the streamkeepers to wrap more pilings under the east dock with weed control cloth in subsequent years. The herring population seemed to grow, but this created a new, albeit in some ways positive, problem: the egg density on the wraps was getting dangerously thick. Thick spawns can cause egg mortality because the eggs at the bottom of the pile can suffocate. The streamkeepers therefore opted to wrap more pilings in weed control cloth to spread out what appeared to be an expanding herring population.

In 2010, the Squamish Harbour Authority, part of the DFO's Small Craft Harbours program, announced the expansion of the Squamish public dock in the Mamquam Blind Channel. The plan was to add more floating wharves and called for new creosote pilings. The streamkeepers had seen how effective the clean black weed control cloth was at attracting spawners and warned the authorities that new creosote might look clean to herring and that they might spawn on it. Their warning went unheeded, and the expansion went ahead as planned.

As the streamkeepers predicted, herring spawned heavily on the new creosote pilings and millions of eggs subsequently died. This event confirmed two things for the streamkeepers: creosote is dangerous for herring eggs, and herring seek out and like to spawn on silt-free surfaces.

The streamkeepers then discovered that herring eggs were also dying on the Squamish Terminals west dock pilings, which were made of concrete. They began to wonder if an ingredient in the concrete was hazardous to herring eggs. They decided to add the Terminals west dock to their project and set about devising a wrap using Velcro and zap straps to hold the material on the concrete pilings without damaging them. This was one of many adjustments the streamkeepers would have to make to their spawning materials as they encountered different wharf and dock designs and environmental conditions.

One of the biggest shifts in their project came when they realized that eggs spawned on pilings would often die from exposure to extreme air temperatures, even if the eggs were protected from creosote by wraps. That told the streamkeepers that they needed to keep the herring eggs submerged. They invented a new spawning surface that looked like a commercial fishing gillnet but had weed control cloth where the net portion would normally be. They called their innovation a float line. It had the usual floats on the top and a lead line on the bottom, so the weed control cloth would stay submerged but could move up and down with the tide. The herring loved the surface, but the streamkeepers discovered another problem. The eggs could be dislodged if waves moved the float line around too much. And if the eggs fell off the float line, they would fall to the silty ocean floor, where they would likely perish. They tried to secure the float line at both ends of the dock, but they could not make it tight enough to counter the wave action. Eggs still fell off the float line.

In 2015, the Squamish Terminals east dock was destroyed by fire. Most of the Squamish Streamkeepers' work was also destroyed in the blaze. At that point the streamkeepers decided to mothball their work at the Squamish Terminals and move on to Fisherman's Wharf in False

Creek. This location presented new challenges for the herring enhancement project. The area is filled with small and medium-sized boats, the wharves are not fixed and so they float up and down with the tides, and there are frequent oil and fuel spills. Even though the streamkeepers wrapped the pilings with a durable plastic wrap to protect herring eggs from the creosote, they soon had to abandon the wraps. They found there were just too many instances of oily substances on the surface of the water, and the eggs laid on the pilings were dying from contact with these contaminants as the tides moved up and down. The streamkeepers then tried a modified float line with a fine mesh netting so it wouldn't get pushed around by tidal currents. Unfortunately, it was too big to manage, and the herring avoided using it because, unlike Squamish where the float line was under a wharf and not exposed to sunlight, here it was exposed to sunlight and consequently became covered in algae. So, they abandoned the wraps and the float line and created a new, smaller spawning surface, which they called a panel. The panel was made from the same netting that was used in the larger float line, but it was only 4 feet (1.2 metres) wide and up to 8 feet (2.4 metres) long. It could easily be tucked around the boats and pilings in a marina. The top of the panel was attached to the wharf, and the mesh, which was weighted by a piece of lead line, sank down into the water. This new panel satisfied all the criteria for the Fisherman's Wharf location and has been used every year since 2016.

As I tracked the streamkeepers' efforts to help the local herring populations, I found myself plagued by a couple of questions: Will these small-scale projects make a difference for Pacific herring populations on the BC coast? And are the herring they are trying to protect endangered populations? It took about 100 years after the arrival of European settlers before herring were exploited commercially for their oil—but after that, it only took several years before herring stocks began to disappear. By the middle of the 20th century, commercial fishing hauls often exceeded 80 percent, and it is believed that this excessive fishing pressure caused the demise of smaller distinct herring populations. The heavy commercial fishing eventually forced the closure of the herring fishery in 1967. This closure was followed by a new mandate to reduce the harvest to about 20 percent of the herring biomass. Currently, the DFO recognizes five major and two minor stocks as defined by their management units. This runs counter to a general belief among much of the interested public that the DFO treats herring on the BC coast as a single, genetically indistinguishable population. Whether this is true or whether the five management units are managed like separate genetic populations, many contend that these large units do not adequately protect minor populations that may still exist within the DFO's large management units.

Canadian government researchers reported in 1941 that the area between Vancouver Island and the mainland was home to many distinct herring populations. It is, therefore, possible that at one time there was a distinct herring population in each and every major inlet on BC's coast. At that time, the researchers observed differences in the number and size of vertebrae among herring collected at different locations. They also believed that intermingling between the stocks was almost non-existent. The DFO, however, may have drawn some of the fuel for its management style from early genetic testing methods that may not have provided enough detail to "separate small-scale stock structures." More recent genetic research on ancient herring bones indicates that the BC coast was once home to at least 43 distinct herring populations. Some people believe that the DFO doesn't want to know more about BC's smaller populations because it could cramp the commercial fishing sector's harvest ambitions. For example, if an endangered population was found to be intermingled with a larger population fishers from the commercial sector were trying to catch, the fishery would have to shut down. As Ken Wilson said during the Cohen Inquiry, one thing that can get politicians and fisheries managers in trouble is to "under-fish." But unless we take a stand and grasp the full details of the distribution, abundance, variability, and migration of each Pacific herring population on the coast, our herring will suffer further damage.

Spawn timing is a key evolutionary driver in creating distinct populations where physical separation is not possible with mobile species like herring. This means that tracking herring spawn timings is a key way for managers to identify, monitor, and protect minor herring populations. The Squamish Streamkeepers have regularly recorded four, possibly five, spring spawn timings in False Creek. This means that distinct populations are still using the Burrard Inlet system, which is remarkable considering how much herring have suffered from living so close to a huge urban area like Vancouver. It also raises a few questions: Are any of the herring spawning in False Creek part of small populations that are in danger of extinction? Or are any of them a segment of a larger population that spawns over a wide area and this is just one spot of many? Or are the five spawns a combination of these scenarios? Which spawns could be considered migratory and which ones are local resident populations? In the last few years, the DFO has been collecting more herring samples and may have a renewed interest in determining the number and health of minor herring populations on the BC coast. It's clear to me now that we have the tools to better understand herring populations—but do we have the will to use them?

In 2013, I got a call from the independent biologist Alexandra Morton. She knew I was working on a film about herring and told me that the herring schooling around the docks at Sointula, BC, were covered in sea lice and were sick. I quickly packed my gear and made a trip up Vancouver Island to try to gather some footage. Morton had played a key role in one of my earlier films, *The Pristine Coast*, when she showed me a photo of a herring bleeding from the creases of its fins. That photo changed that film from a story about open net pen fish farms' impacts on wild salmon to their impacts on all wild fish populations. Since that time, I've recorded heavy sea lice loads on juvenile herring at Campbell River, BC, and then, to my dismay, at False Creek. The streamkeepers have been trying incredibly hard to bring back the herring at False Creek, so it was more than disappointing to discover them sick and covered in sea lice.

The Canadian government launched an inquiry into the decline of the Fraser River sockeye populations in 2009. I filmed a good portion of the evidentiary hearings in 2010 and 2011 and posted the footage on The Green Channel under the title *Exhibit 2148*. I created two films from the hearings: *The Unofficial Trial of Alexandra Morton* and *Trial of an Iconic Species*. The latter includes clips of testimony by Dr. Gary Marty, who worked for the Province of British Columbia as a fish pathologist. He stated that "viral hemorrhagic septicemia virus [vHSv] is the most common identified cause of these lesions of concern." He was talking about the most common virus behind lesions on fish that were sent to his lab for evaluation. This statement piqued my interest because I was already researching the concerns Canadian bureaucrats had in the 1980s about this virus and its likely introduction to BC marine waters if we were to import Atlantic salmon eggs from Europe to stock open net pen fish farms. As it turns out, Atlantic salmon eggs were imported to North America in the late 1970s, first to US fish farms in Puget Sound and then to Canadian ones. It was a bad decision by both countries: vHS and other diseases were introduced and soon began to infect wild fish populations up and down North America's Pacific coast. Disastrous population declines of most, if not all, fin fish species including herring have been recorded since the appearance of vHS and other exotic pathogens.

Dr. Marty's team led an extensive study of the *Exxon Valdez* oil spill that occurred in Prince William Sound on March 24, 1989, to determine its impact on Pacific herring populations. Alaskan herring populations crashed shortly after that spill, and the thinking at the time was that the oil spill was the reason. Dr. Marty, however, determined that herring could easily metabolize oil, which led the researchers to look for other causes. He found that viral hemorrhagic septicemia (vHS) was prevalent at the population level and determined that it was the cause of the herring population crash. We don't know if vHS was already in the herring populations before the spill occurred or if it just happened to arrive shortly after the spill. But the evidence does seem to indicate that it arrived somewhere around 1989. This means that VHS took about 20 years to move up the Pacific coast from Washington State to Alaska.

The bad news for herring regarding diseases doesn't end with VHS, though. When I interviewed Dr. Gideon Mordecai for *The Herring People*, he explained that another virus, erythrocytic necrosis virus (ENV), was impacting Pacific herring. This virus makes fish anemic, and herring are particularly susceptible to it. The main focus of his research has been endangered wild salmon, and he discovered a number of new viruses infecting them that were traced back to open net pen farms. He suggested that if his research team were to conduct similar tests on herring, they would likely also find a number of previously undiscovered viruses. The potential impact of those viruses, if they exist, of course, is unknown. But it makes one wonder what impact all these viruses are having on the marine ecosystem beyond what we already know. For example, the ENV variant found in herring is almost identical to the variant found in chinook salmon. Given the strong predator-prey relationship between chinook salmon and herring, it's reasonable to assume that herring are infecting chinook with this virus and contributing to the decline of chinook populations. If that is in fact the case, it has implications for our endangered Southern Resident killer whale population, because chinook salmon is its main food source. Is the introduction of open net pen fish farms the real culprit in the decline of our Southern Resident killer whale population?

It's one thing to get a full understanding of how these diseases are affecting each individual herring and its survival. It's a whole other thing to grasp how the spread of these diseases will move through herring populations and impact all the other fish species that mingle with and feed on herring at different stages of their life cycle. For example, during the making of *The Herring People,* I filmed three-spined stickle-backs (*Gasterosteus aculeatus*) feeding on herring eggs at Fisherman's Wharf. A week later, I noticed that many of the sticklebacks had mucus trailing from their vents. Although there hasn't been any testing to investigate the cause of the mucus in this case, there is plenty of literature that says it is a sign of infection. So what infection did the sticklebacks acquire, and did they get it from eating herring eggs? Was it lethal to them? Did they spread it to other species of fish? My stickleback discovery is just a small example of how connected the marine food web is and how quickly a disease could spread through it.

At the end of *The Herring People*, I suggest that five major human activities, which I call "cuts," have impacted herring since settlers arrived on the Pacific coast of North America: (1) overharvesting during the Reduction Fishery Era, which likely reduced herring population diversity; (2) over-exploitation of resident or local herring stocks; (3) spawning habitat degradation from shoreline development in areas frequented by herring spawners; (4) damage to herring eggs from oil or oily contaminants; and (5) the introduction and amplification of lethal diseases from open net pen fish farms. It's clear we have a lot of work to do to protect Pacific herring on British Columbia's coast. I hope the takeaway from *The Herring People* is that herring are a keystone species in the North Pacific, and that if they are in trouble, so is the entire marine ecosystem.

The Squamish Streamkeepers have been trying to help the herring of Howe Sound and False Creek since 2006. The group now has a second generation of dedicated volunteers who want herring populations to return to their former levels of abundance. These projects may seem too small in scale to make a real difference. But I believe the publicity surrounding them has changed attitudes and increased awareness about herring spawning habitat. In Squamish, the Mamquam Blind Channel is undergoing redevelopment and transitioning away from industrial activities to residential developments. Even though this type of development should reduce pollution in the Mamquam Blind Channel on its own, during the planning stages there was much discussion about how to avoid impacting herring spawning activity.

New wharves or docks have been constructed to replace creosote pilings up and down the coast. And when new concrete pilings were placed at the Squamish Terminals, the people involved drew on lessons learned from the Squamish Streamkeepers' activities and put a shiny surface on them to discourage herring from spawning on them. So the streamkeepers have unarguably made a difference. Their projects brought a lot of attention to the plight of herring, which will, in the end, benefit herring populations up and down the Pacific coast.

The sad news that Woody Morrison, a key contributor to *The Herring People*, passed away on January 28, 2021, still weighs heavy in my heart. His beautiful words about the herring people as described in Haida Nation oral history sum up what herring should mean to all of us. I chose the title *The Herring People* because it also describes all the people, many of whom appear in my film, who protect, manage, study, harvest, admire, and consume this amazing and vital species. We need to respect this species, like First Nations communities do, for the important role it plays in the marine food web off of BC's coast and protect it not only for our own sake, but for the sake of all species that rely on it.

Cast of Characters

Pacific Herring:
   Clupea pallasii

Woody Morrison:
   Storyteller and Historian, Haida Nation

John McLachlan:
   Singer

Dr. Douglas Hay:
   Scientist Emeritus, Fisheries and Oceans Canada

Percy Redford:
   Commercial Fisherman, Herring Roe-on-Kelp Fishery

Grant Scott:
   Local Trustee, Hornby Island

Calvin Siider:
   Commercial Fisherman, Herring Sac-Roe Fishery

David Ellis:
   Former Head, Pacific Fisheries, COSEWIC

Dr. Thomas E. Reimchen:
   Biology Professor, University of Victoria, BC

Dr. Jonn Matsen:
   Herring Coordinator, Squamish Streamkeepers

John Buchanan:
    Conservationist, Squamish, BC

Eric Andersen:
    Local Historian, Squamish, BC

Edith Tobe:
    Executive Director, Squamish Watershed Society

Douglas Swanston:
    Biologist, NW Seacology

Fred Felleman:
    Environmental Consultant, Seattle, Washington

Kurt Stick:
    Biologist, Washington Department of Fish and Wildlife

Dr. Jeff Marliave:
    VP Marine Science, Vancouver Aquarium

Richard Steiner:
    Retired Professor, University of Alaska

Dr. Gary Marty:
    Research Associate, University of California, Davis

Dr. Gideon Mordecai:
    Research Associate, Institute of Oceans and Fisheries, UBC

The Herring People
—an illustrated screenplay

By
Scott Renyard

JUGGERNAUT PICTURES LOGO

                                    FADE OUT:

FADE IN:

INT. MAMQUAM BLIND CHANNEL — DAY

Eerie music. It's a spooky underwater
environment. Brown pillars sit in murky
water. It's like an underwater city. Has the
world flooded?

AT THE BOTTOM — are cinder blocks that look
like the remains of cement stairs and, oddly,
a sheet of cloth covered in silt.

PANNING ACROSS — a solid wall. There is
finally a sign of life, small patches of
slimy green algae.

                    WOODY MORRISON (V.O.)
                It was during the Hunger Time,
                and this boy was always hungry.
                He wanted something to eat. His
                mother gave him these smoked
                collars from the salmon, and
                there was a little bit of mould
                on one. He didn't want it.

CLOSER — on the pillars. They are covered in
muck and brown slime. We see now that they
are pilings.

>   WOODY MORRISON (V.O.)
>   His mother told him to take it
>   down to the ocean and wash it
>   in the sea water and it'll be
>   good.

RUSTY CABLES — hang along the wall, obviously
not in use.

MORE ALGAE — but not much, as barely anything
can survive in this compromised environment.

>   WOODY MORRISON (V.O.)
>   His guide was a woman who was
>   half-stone. She'd violated some
>   major taboo called Anan Nam.

SOME CINDER BLOCKS — are stacked on top of a
big block of concrete. Was this an old house?
Was it flooded?

>   WOODY MORRISON (V.O.)
>   And she said, "You go over to
>   the house. There's an opening
>   in the corner. You hang it over
>   the wall and hold on to it.
>   Don't let go, but don't look,
>   don't look in there."

ANOTHER WALL — appears, but it's different from the previous one. Not wood pilings, but perhaps a row of concrete pilings. They are covered in dead barnacles, which are in turn smothered in brown muck.

>   WOODY MORRISON (V.O.)
>   One day he heard all this
>   merry-making.

THE BARNACLES — are just skeletons of what was once a vibrant ecosystem.

A DARK HOLE — appears between some rotting pilings.

>   WOODY MORRISON (V.O.)
>   He was curious, so he peeked
>   through a hole and he got
>   herring roe in his eye.

MOVING UP — the wall. At the top, there is a bit of light. Just a glimpse of the world above.

>   WOODY MORRISON (V.O.)
>   So he asked, "What's going on?"
>   "Oh, that's the herring people."
>   "Who are the herring people?"
>   "Oh," she says, "they're very
>   important people, there's many
>   of them."

FADE OUT:

FADE IN:

EXT. SQUAMISH ESTUARY (AERIAL) — DAY

Looking down at a compromised estuary.
Logging dumps and a shipping terminal
dominate the land next to the water.

A LOG SORT — on the edge of the water is
active, with an operator moving a log with
his machine.

EXT. CITY OF VANCOUVER (AERIAL) — DAY

False Creek and the Burrard Bridge are in the
foreground. Numerous boats are moored at a
large wharf network that covers most of the
water's surface.

A LONE MAN — walks along a float that is part
of Fisherman's Wharf. It's Jonn Matsen.

EXT. MAMQUAM BLIND CHANNEL (DRONE) — DAY

Flying low across the water, we see
not all of the Mamquam Blind Channel is
industrialized. But this is definitely not a
place untouched by human activity.

TITLE: Juggernaut Pictures Presents

LOOKING BACK — at the Mamquam Blind Channel,
we see log booms cover a large part of it.

A MAN — has left his boom boat and is walking on the raft of logs.

STRAIGHT DOWN — a bird's eye view of bundles of logs tied with cables.

EXT. WEST COAST (DRONE) — DAY

A spectacular view of a British Columbia west coast island. The water in the foreground is white with the milt of herring spawning. It's Hornby Island.

TITLE: A Scott Renyard Documentary

ALONG A SHORELINE (DRONE) — A small commercial fishing boat sits in a patch of milty water.

> JOHN MCLACHLAN (V.O.)
> From long before your time,
> from long before your day, I've
> roamed the oceans wide.

STRAIGHT DOWN (DRONE) — along a shoreline covered with a thick layer of herring milt. Gulls dot the shoreline and several of them take flight over the water. Sea lions emerge from and disappear back into the milty water.

> JOHN MCLACHLAN (V.O.)
> Not knowing where I'm bound,
> but knowing where to go,

TITLE: In association with Pacific Coast Entertainment Ltd.

ALONG A SHORE (DRONE) — Not much milt this time, but we can see an impressive school of herring spawners, presumably looking for a place to spawn.

> JOHN MCLACHLAN (V.O.)
> . . . I return to this place.

EXT. HORNBY ISLAND (DRONE) — MOMENTS LATER

A few gillnetters work the shoreline near milt-laden water.

> JOHN MCLACHLAN (V.O.)
> And the shallows and the weeds
> in the seething silky seas, a
> whole universe explodes.

PANNING — a tip of Hornby Island. The water is milty here as well. A spectacular sunset reflects off the water.

UNDERWATER — A large school of herring spawners swim through mildly milty water.

> JOHN MCLACHLAN (V.O.)
> New life it is spun,

FROM ABOVE (DRONE) — A small raft of sea lions swim through clean waters.

                    JOHN MCLACHLAN (V.O.)
          . . . new life makes its way,

EXT. CHANNEL — DAY

Sea lions bob in the water. Then a small
school of herring enter the frame and splash
like rain hitting the water.

                    JOHN MCLACHLAN (V.O.)
          . . . in spring I scatter like
          the rain.

BACK UNDERWATER — The herring swim by close
to camera. But something isn't right. Large
salmon are lurking behind them. Lots of them.

                    JOHN MCLACHLAN (V.O.)
          I am the herring, help me 'fore
          I'm gone.

EXT. INDUSTRIAL-SIZED FISH FARM — DAY

Gliding along the water, equipment is poised
alongside the pens, ready to deposit fish
into the pen.

                    JOHN MCLACHLAN (V.O.)
          I am the herring, help me 'fore
          I'm gone.

INT. OCEAN — DAY

This time the herring are inside the fish
farm and swimming with the Atlantic farm
fish.

TIGHTER — Herring swim very close together
through green water.

INT. OCEAN — DAY

Tracking over submerged riprap, we see the
rock is covered in a white coating. It's
fresh herring eggs.

TITLE: The Herring People

                            FADE OUT:

FADE IN:

EXT. MOUNTAIN RANGE — DUSK

A full moon sits in a clear sky over snow-
capped mountains. It looks very cold.

               WOODY MORRISON (O.S.)
        During the ice age,

INT. WOODY MORRISON'S OFFICE — DAY

Establish. Interview.

Woody Morrison, a Haida Elder, is in his
office.

LOWER THIRD: Woody Morrison, Storyteller and Historian, Haida Nation.

> WOODY MORRISON
> . . . the winters were getting
> longer, and the spring was
> taking longer, and so it became
> known as the Hunger Time.

EXT. SKY — DAY

A pale sun barely peeks through the high cloud cover. A bit of snow covers the hills.

ANOTHER SNOW-CAPPED MOUNTAIN — Snow begins to fall.

PANNING ACROSS — another mountain range. Now it's snowing heavily.

> WOODY MORRISON (O.S.)
> During that period is when we
> have to depend on the food that
> we've put up in the fall time.

SNOW — falls heavily into the ocean and onto a small west coast island.

TREES — are covered in snow. More snow is falling.

> WOODY MORRISON (O.S.)
> And during one of those times
> they were running out of food.

CLOSER — It's snowing so heavily it's difficult to see the branches.

ON A BARREN HILL — Grasses are barely visible above the snow and the wind whips up blizzard-like conditions.

WIND — blows snow through a frozen estuary.

> WOODY MORRISON (O.S.)
> So they start cutting back on how much they're eating, then they start getting weak, but there's still no spring.

EXT. SHORELINE — DAY

Herring are flipping around in shallow water in a frenzy of spawning activity. An eagle is eating one. It's clear they are very accessible to humans when they are so close to shore.

> WOODY MORRISON (O.S.)
> The herring are very, very important, and the herring are one of the first ones that come in.

FADE TO:

MAP — of the Danish coast.

YELLOW DOTS — spread across the coastline to mark the locations of middens that contain herring bones.

> NARRATOR (V.O.)
> Archeological studies of human settlements on the Danish coast revealed that herring played a key role in the diets of Europeans dating back 5,000 years.

MAP — of the British Columbia coast.

YELLOW DOTS — mark herring bone findings from Alaska to the Canada-US border.

> NARRATOR (V.O.)
> A more recent 2013 study found herring bones at 171 ancient human settlements from Alaska to Washington State.

INT. OCEAN — DAY

A school of spawning herring occupy the green water.

> NARRATOR (V.O.)
> This means that herring were important to humans right across the Northern Hemisphere and . . .

EXT. ESTUARY SHORELINE — DAY

Looking straight down, herring can be seen swimming through eel grass.

>   NARRATOR (V.O.)
>   . . . it extends herring's role
>   in human diets to at least
>   10,000 years.

UNDERWATER — A small school of herring twist and turn in the water as they feed on something too small for us to see.

                                    FADE TO:

OLD MAP — of the British Columbia south coast.

ANIMATED — dotted red line enters the Salish Sea, tracking the path taken by European settlers when they arrived by ship at this part of North America.

>   NARRATOR (V.O.)
>   Less than 250 years ago,
>   European settlers began
>   arriving on the shores of
>   Canada's Pacific coast, led by
>   Juan Pérez Hernández.

A PAINTING — of the HMS *Endeavour*, James Cook's ship.

European settlers, led by Juan Pérez Hernández, arrived on the BC coast in 1774. James Cook arrived in 1778 on his ship the HMS *Endeavour*. This photo shows early ships moored in Burrard Inlet. (Source: City of Vancouver Archives)

                    NARRATOR (V.O.)
          And just four years later,
          James Cook arrived looking for
          territories useful to England.

PHOTO — of Burrard Inlet with a collection of
sailing ships moored in English Bay.

PHOTO — of early settlers who built a house
on the shores of Burrard Inlet.

                    NARRATOR (V.O.)
          The resource-rich North Pacific
          was useful, and they began to
          put down roots.

PHOTO — of a settler in a row boat near
Stanley Park.

PHOTO — of small fishing boats near the mouth
of the Fraser River.

                    NARRATOR (V.O.)
          It would take nearly a century,
          however, before the new
          settlers would target herring
          in a big way.

PHOTO — of the growing city of Vancouver,
late 1800s.

PHOTO — of a small fleet of herring boats in
English Bay.

                    NARRATOR (V.O.)
          Records indicate that in 1877,
          herring were harvested for a
          Vancouver oil-based extraction
          business.

PHOTO — of the Coal Harbour area. The
downtown area is full of smoke from fires.

PHOTO — of a factory fire. A sign displaying
the words "The Canadian Fish and Cold Storage
Company" is on a neighbouring building.

PHOTO — of firefighters battling to save a
building on fire.

                    NARRATOR (V.O.)
          But just nine years later, in
          1886, the operation was shut
          down by a factory fire,

DOCUMENT — BC Fisheries Report, page 109.

PUSHING IN ON THE TEXT — the words "largely
left Burrard Inlet" lift off the page.

                    NARRATOR (V.O.)
          . . . and was not reopened
          because, in the words of a
          Fisheries inspector, herring
          had "largely left Burrard
          Inlet."

ANOTHER PAGE — BC Fisheries Report, page 110.

This photo is taken from the south shore of False Creek, looking west to Burrard Inlet. On the left is Senákw, which is now known as Kitsilano Point. It was the winter village of the Squamish Nation. (Source: City of Vancouver Archives)

Early fishing boats were small but effective at catching herring in the waters in and around Burrard Inlet, near Vancouver. It took about 100 years before the settlers began to commercially exploit herring—and then it took only about nine years for the herring stocks in Burrard Inlet to be so depleted they could no longer support a commercial fishery. (Source: City of Vancouver Archives)

A fire like this one in 1886 shut down the herring oil extraction business. The business didn't reopen after the fire because the local herring stocks had been overfished. (Source: City of Vancouver Archives)

THE TEXT DARKENS — The date, "1892," lifts off the page, then panning down, the words "Destroying" and "immense quantities" lift off the page separately. Then the words "a useless waste" join them.

> NARRATOR (V.O.)
> And the Fisheries Commission determined in 1892 that destroying "immense quantities" of herring for their oil was "a useless waste."

INT. OCEAN (1892) (RE-ENACTMENT) — DAY

A large school of herring swim through a spooky kelp forest.

EXT. BURRARD INLET SHORELINE (1892) (RE-ENACTMENT) — DAY

A large flock of seabirds gather near the shore in search of food.

> NARRATOR (V.O.)
> And by then, herring were disappearing from inlets and bays up and down the Pacific coast.

MORE BIRDS — fly around a seiner.

HERRING — in a seine net struggle in vain to escape as they are scooped out of the ocean.

PHOTO — of a shipping dock on an early part of industrialized Vancouver waterfront.

> NARRATOR (V.O.)
> Catches continued to grow
> during the first three decades
> of the 20th century, when most
> of it was exported to Asia for
> food.

DOCUMENT — Department of Fisheries. (1937). Seventh annual report of the Department of Fisheries: For the year 1936-37.

PANNING DOWN — the cover to the words "For the year, 1936-37."

PANNING OVER TO PAGE 11 — finding the word "Herring" then over to "162,062,500 pounds" and down to "61,211,800" pounds.

> NARRATOR (V.O.)
> In 1936, the herring catch
> coast-wide weighed over 162
> million pounds, an increase
> of 61 million pounds from the
> previous year.

ANOTHER DOCUMENT — Fisheries Research Board. Early results of individual herring stocks.

ZOOMING IN — on the words "150,000 tons," which morph into "240 million pounds."

                    NARRATOR (V.O.)
          And reaching "240 million
          pounds" in 1941.

DOCUMENT — Hourston, A.S. (1978). Fisheries
and Marine Service Technical Report 784.

PANNING — to find the words "concern for
conservation." They lift off the page.

BRITISH COLUMBIA MAP — Zones are defined by
lines. The South Coast zone is highlighted
in yellow with the text "1936" in black,
the Middle Coast zones are highlighted in
orange with the year "1940" in black and the
last zones, Central and Northern Coast, are
highlighted in red and joined by the year
"1941."

                    NARRATOR (V.O.)
          Fishing pressure was far too
          great, and the "concern for
          conservation" forced the
          Canadian Government to set
          catch quotas on the south coast
          in 1936, the middle coast in
          1940 and the north and central
          coast in 1941.

                                    FADE TO:

PHOTO — of a commercial seiner pulling in a
net full of herring.

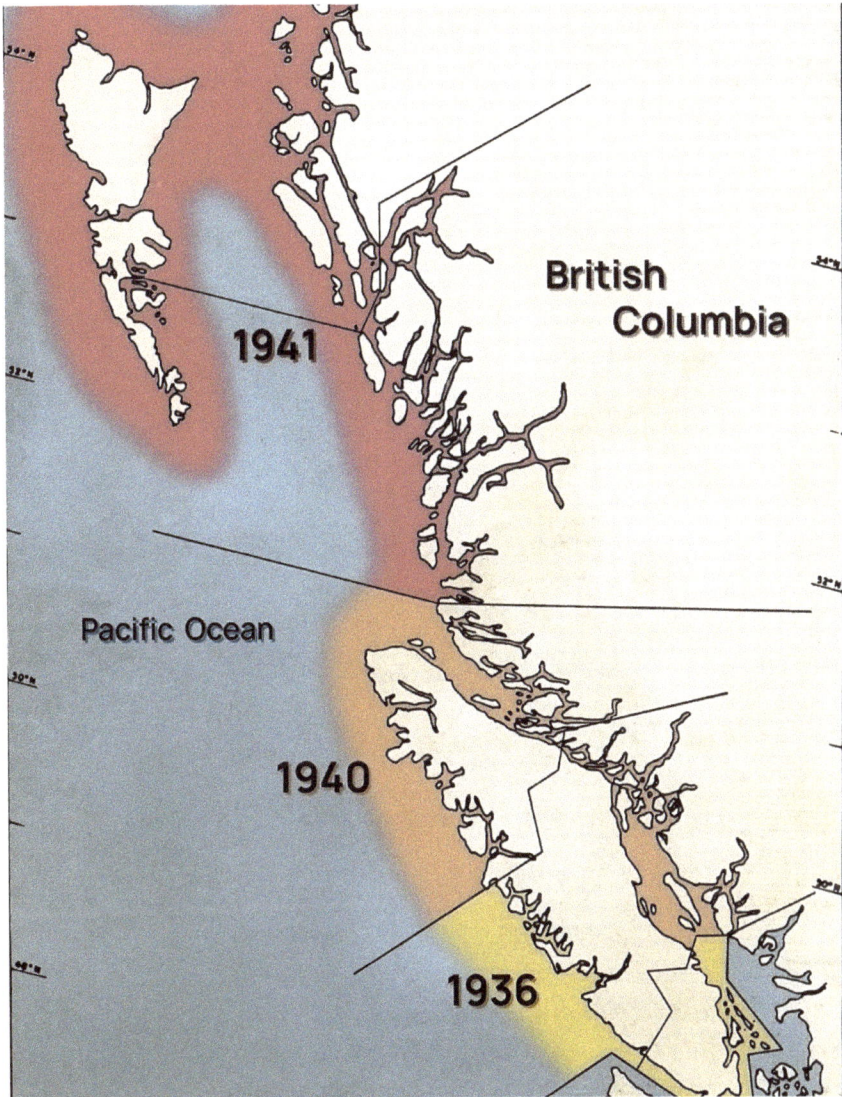

By the mid-1930s, fishing pressure forced the government to start setting catch quotas on the BC coast. (Source: Taylor, F.H.C. (1964). Life history and present status of British Columbia herring stocks. Modified by Pacific Coast Entertainment Ltd.)

PHOTO — of a fisherman watching while a scoop of herring are brailed out of the seine net.

PHOTO — of fishermen guiding a basket of fish over the deck of their boat.

> NARRATOR (V.O.)
> By the end of the Second World War, herring's value as whole fish dropped, pushing the market to turn herring into fish oil and fish meal products. This was known as the Reduction Fishery Era.

EXT. HERRING FISHING GROUNDS (AERIAL) — DAY

Commercial fishing boats are scattered over the ocean below.

EXT. OCEAN — DAY

Commercial fishing boats work in a moody ocean.

> NARRATOR (V.O.)
> But even with quotas, the catches on the Pacific coast were still too large, and the stocks crashed, forcing the closure of the herring fishery in 1967.

Fishing pressure continued to grow, even with catch quotas in place, which forced the closure of the herring fishery in 1967. It was common to catch 1,000 or more tons in a single set, which put tremendous pressure on herring populations. (Photo: Fisherman Publishing Society)

EXT. LARGE SEINER — DAY

Looking down over the side of a large seiner.
A net is full of herring.

EXT. FISHING GROUNDS (1967) (RE-ENACTMENT) —
DAY

A fisherman on a small gillnetter pulls his
net over the back of the boat. The net is
empty.

           DOUG HAY (O.S.)
     After the years of the
     Reduction Fishery,

EXT. BEACH — DAY

Establish. Interview.

Doug Hay, a retired DFO scientist who
specialized in herring, is sitting on a beach
chair.

**LOWER THIRD: Dr. Doug Hay, Scientist
Emeritus, Department of Fisheries and Oceans
Canada.**

           DOUG HAY
     . . . where sometimes in excess
     of 50 percent of the fish
     were captured every year, um,
     people didn't want to see that
     repeated.

EXT. SEINER — DAY

A seine net full of herring is pulled toward the back of the boat.

>   NARRATOR (V.O.)
>   Some researchers contend that
>   the capture rate had often
>   reached as high as 80 to 90
>   percent, which . . .

TIGHTER — on the net full of herring.

VERY CLOSE — The small fish struggle in vain.

>   NARRATOR (V.O.)
>   . . . would have likely caused
>   permanent damage to some
>   herring populations.

OVER THE SIDE — of another seiner. The net is loaded with herring.

RESUME — Hay interview.

>   DOUG HAY
>   When you look around the world,
>   you see that almost every
>   population of herring at one
>   time or another has crashed.
>   But invariably they come back.

ANIMATED HERRING GRAPH — of the British Columbia herring catch since 1900.

The catch grew until the late 1960s and then plunged to a very low level.

> DOUG HAY (O.S.)
> They collapsed in 1966-67. By 1970 they had re-established.

UNDERWATER — Herring fill the frame as a school swims through deep water.

> DOUG HAY (O.S.)
> Sometimes it takes a year or two or three as they did in British Columbia.

EXT. JAPANESE MARINA — DAY

A small commercial fishing boat cruises through the marina, heading to the open ocean.

ON THE FISHING GROUNDS — A fisherman pulls in his gillnet.

FROM ANOTHER BOAT — The fisherman continues to pull in his net.

> NARRATOR (V.O.)
> Across the Pacific Ocean, the Japanese herring stocks also crashed in the late 1960s.

TIGHT — on the fisherman guiding his net into the boat.

INT. OCEAN — DAY

Herring swim between fronds of sugar kelp
that are covered in herring eggs.

> NARRATOR (V.O.)
> These shortages in Japanese roe
> supply . . .

HERRING — swim slowly through some kelp.

> NARRATOR (V.O.)
> . . . created an opportunity
> for Canadian products.

TIGHTER — Herring are milling around kelp
fronds that are already covered with herring
eggs. The herring are actively spawning.

DOCUMENT — MacLeod, J.R. (1972). The herring
fishery: Potential for expansion.

THE PAGE DARKENS — and the words "Report of
the Herring Task Force 1972" lift off the
page as a piece of torn paper.

> NARRATOR (V.O.)
> A Herring Task Force was
> formed in 1972 by the Canadian
> Government to evaluate the
> Pacific Coast Herring Fishery.

PANNING DOWN — and finding the word "achieve"
then over to the words "$100 million sales
target, possibly by 1980."

> NARRATOR (V.O.)
> Its mandate was to see if
> the herring resource could
> "achieve" a "$100 million sales
> target, possibly by 1980."

INT. PERCY REDFORD'S HOUSE (GREEN SCREEN) —
DAY

Establish. Interview.

Green screen plate: A beautiful shoreline.

**LOWER THIRD: Percy Redford, Commercial
Fisherman, Herring Roe-on-Kelp Fishery.**

> PERCY REDFORD
> I got my licence with licence
> 24. The 24th licence.

EXT. ROE-ON-KELP POND WHARF — DAY

A young Percy Redford with family and staff
harvesting their roe-on-kelp on a float next
to their herring pond.

Percy Redford, a commercial roe-on-kelp fisherman, and his crew harvesting herring roe that was spawned on the sugar kelp fronds in their ponds. The roe-on-kelp are trimmed onsite and the trimmings are returned to the pond so the eggs on them can hatch out. The spawners are released back into the wild and, if the release is done properly, will survive to spawn again. (Source: Dick Harvey)

                    PERCY REDFORD (O.S.)
          I was operational on the
          second year of the licensing.
          So, if you had a little bit of
          knowledge, you got a licence.

ANOTHER ANGLE — of the pond operation. They
slice up the kelp with herring roe and place
the product in large totes.

ON A KNIFE — Percy slices a frond and tosses
the trimming to the side.

EXT. BAY — DAY

A seiner has its net out in the standard
loop, holding herring in a makeshift pond.

                    PERCY REDFORD (O.S.)
          And a lot of the seiners and
          the people that got into it,

IN THE NET — The herring swim in a circle
inside the seine net.

                    PERCY REDFORD (O.S.)
          . . . didn't really have the
          knowledge to know what they
          had in the net. And they would
          overcrowd the fish . . .

RESUME — Redford interview.

>   PERCY REDFORD
>   . . . and this is where most of
>   the fish were dying from.

INT. OCEAN — MOMENTS LATER

A large school of herring are crowded inside
the net. They are swimming frantically.

>   PERCY REDFORD (O.S.)
>   And then if they didn't spawn
>   in three days, they would let
>   them go.

FROM THE SIDE — The herring head deeper into
the net and then out the opening.

>   PERCY REDFORD (O.S.)
>   These fish would swim off in
>   different directions, they
>   wouldn't be in a school
>   anymore . . .

EXT. OCEAN — MOMENTS LATER

Sea lions sitting outside the net grab the
escaping herring.

>   PERCY REDFORD (O.S.)
>   . . . and the predators would
>   eat them up pretty quickly.

RESUME — Redford interview.

                    PERCY REDFORD
          We knew that the density had
          to be just right or else the
          herring would die.

UNDERWATER — The herring swim by in a dense
school.

                    PERCY REDFORD (O.S.)
          So my system was to get the
          fish that were ready to spawn
          off the beach in shallow water
          where the seine boat couldn't
          get.

EXT. BAY — EARLIER

A seiner is catching herring in its net in
deeper water.

LOOKING DOWN — into the water. Herring are
pushed up against the net.

UNDERWATER — Redford's herring are now
swimming through the kelp in his pond. The
kelp has been strung out on lines.

                    PERCY REDFORD (O.S.)
          I mean, some of these fish
          were taken off in six to eight
          inches of water and they would
          hit the lead, come back into
          my pond and then be released
          into the seaweed and they would
          spawn immediately.

RESUME — Redford interview.

>   PERCY REDFORD
>   And it was a good system until
>   the sea otters showed up.

EXT. KELP BED (RE-ENACTMENT) — DAY

A raft of sea otters rest in a kelp bed.

>   PERCY REDFORD (O.S.)
>   We had one sea otter come
>   in and he would swim into my
>   gear and he would kick a . . .

A SINGLE SEA OTTER — rolls in the surf with a
stipe of kelp in its paws.

>   PERCY REDFORD (O.S.)
>   . . . stipe of kelp off, which
>   is about 12-14 leaves on a
>   stem, and he would go and bite
>   it, and he would take the whole
>   stipe out into the middle of
>   the bay and then he would sit
>   out there and eat it. He would
>   grab it in his front legs and
>   he would chew on it until it
>   was gone.

RESUME — Redford interview.

> PERCY REDFORD
> Then he would go back for
> another, and he only took the
> best, never took the worst.

EXT. SHORELINE (RE-ENACTMENT) — DAY

A few sea otters frolic along the shoreline.
Two of them grab and eat a herring.

> PERCY REDFORD (O.S.)
> And we can't control the sea
> otters. It's the end of the
> game as far as I'm concerned.

RESUME UNDERWATER — The spawners mingle
around the kelp set out in the pond.

> NARRATOR (V.O.)
> In spite of the problems,

A FEW HERRING — swim between the kelp blades.

> NARRATOR (V.O.)
> . . . the roe-on-kelp fishery
> is thought to have a lower
> impact on herring populations
> because . . .

A LARGE SCHOOL — swim through an opening and
are released back into the wild.

NARRATOR (V.O.)
. . . the egg harvest is
controlled and, in theory, the
spawners are returned to the
wild to spawn again.

INT. OCEAN — MOMENTS LATER

A school of herring change direction and
disappear down into the spooky depths of the
ocean.

FADE TO:

EXT. DENMAN ISLAND (DRONE) — DAY

A fleet of commercial fishing boats are
spread out along the shore, stretching almost
to the horizon.

NARRATOR (V.O.)
The sac-roe fishery, however,
continues to be mired in
controversy.

FROM ABOVE (DRONE) — The deck of the herring
skiff is covered in herring and the fishermen
are shovelling them into the hold.

INT. GRANT SCOTT'S BOAT — DAY

Establish. Interview.

LOWER THIRD: Grant Scott, Local Trustee, Hornby Island.

> GRANT SCOTT
> Sixty-eight thousand people
> signed a petition we put out
> there. We thought maybe we'd
> get a thousand. That would
> be huge! Sixty-eight thousand
> people so far have signed this
> thing. Why?

A GILLNETTER — is pulling in its net.

> GRANT SCOTT (O.S.)
> Because the people can see the
> environment around them and
> they're concerned.

ON THE NET — Herring spawners are dragged out
of the water at one end of the skiff and into
the boat.

> GRANT SCOTT (O.S.)
> We started to look at the
> statistics . . .

THE DECK — is covered in herring spawners.

> GRANT SCOTT (O.S.)
> . . . and what's actually
> happening with these fish that
> are around our island.

VERY CLOSE — The herring are being dragged
and shaken across the deck.

>   GRANT SCOTT (O.S.)
>   We just came to this point of
>   saying, you know, we've gotta
>   find a better way to manage
>   this whole thing because . . .

UNDERWATER — Herring swim past some bright
green algae.

>   GRANT SCOTT (O.S.)
>   . . . the herring are such a
>   foundation fish for the whole
>   Strait of Georgia ecosystem
>   and the ecosystem around Hornby
>   Island.

EXT. OCEAN (DRONE) — DAY

A cluster of commercial fishing boats are
tied up together in a holding pattern.

ON A SKIFF — It's floating in the middle of a
milt-laden surf and pulling in a full net. A
large paddle knocks the herring violently out
of the net.

>   NARRATOR (V.O.)
>   After catching too many fish
>   prior to 1970,

EXT. SHORELINE — LATER

Fishing vessels criss-cross in choppy water.

ON A GILLNETTER — pulling in its net. There are very few fish.

THEN CLOSER — This part of the net is much fuller.

> NARRATOR (V.O.)
> Canadian fisheries managers
> set the Pacific herring harvest
> at 20 percent of the estimated
> biomass. And for a few years
> the strategy worked.

HERRING — dart through some eel grass.

> NARRATOR (V.O.)
> Herring began to recover.

EXT. FORD COVE WHARF — DAY

Establish. Interview. Calvin Siider is sitting on the back of his boat.

LOWER THIRD: Calvin Siider, Commercial Fisherman, Herring Sac-Roe Fishery.

RESUME — A gillnetter pulling in a full net.

Workers in a sac-roe processing plant remove the roe and package it into crates for shipping. Sac-roe can be seen floating in the totes in the foreground. It is mostly exported to Japan. The rest of the fish is ground up and turned into fish meal or fish oil products. (Photo: Doug Hay)

> CALVIN SIIDER
> Since we started fishing this
> stock of herring in the Gulf
> of Georgia it has been on the
> incline for at least 20 years.

EXT. LAMBERT CHANNEL — DAY

A gillnetter hauls in a loaded net.

> CALVIN SIIDER (O.S.)
> If we were hurting this stock
> by taking 20 percent,

UNDERWATER — Herring are caught in the
gillnet.

> CALVIN SIIDER (O.S.)
> . . . it would be on the
> decline.

RESUME HERRING GRAPH — The increase in catch
seen in the 1970s stalls. The catch drops and
stays down until the present day.

> NARRATOR (V.O.)
> But something was still wrong.
> After a decade or so, herring
> populations began to decline
> again.

> DAVID ELLIS (O.S.)
> As I grew up . . .

INT. LIVING ROOM (GREEN SCREEN) — DAY

Establish. Interview.

Green screen plate: Ford Cove marina, Hornby Island.

LOWER THIRD: David Ellis, MSc, Former Head, Pacific Fisheries, COSEWIC (Committee on the Status of Endangered Wildlife in Canada).

> DAVID ELLIS
> . . . I was a sport fisherman around the Strait of Georgia and watched that dry up in my lifetime to . . .

EXT. DENMAN ISLAND SHORELINE — DAY

A sport fisherman jigs for herring out of his boat.

> DAVID ELLIS (O.S.)
> . . . a fraction of what it used to be.

UNDERWATER — A school of salmon swim in a school through the open ocean.

> DAVID ELLIS (O.S.)
> Probably five percent of the chinook and coho still exists.

EXT. BAY — DAY

A seiner drifts along, empty, with its nets on the drum. No fish.

> DAVID ELLIS (O.S.)
> Catches used to be phenomenal.

ON A SEINE NET — It's being pulled up. This time there are a few fish.

> DAVID ELLIS (O.S.)
> I'm all for lots of fishermen. But they have to be designed to catch surplus and not bite into the capital.

ON THE FISHING GROUNDS — Gillnetters are working Lambert Channel.

ANOTHER BOAT — The crew drags the catch over the back of the boat.

> NARRATOR (V.O.)
> After more than 30 years of herring declines, the Canadian Government was forced to cut the harvest to just 10 percent of the estimated biomass . . .

A COMMERCIAL BOAT — is dragging its skiff to another location.

                    NARRATOR (V.O.)
          . . . in a last-ditch effort to
          save British Columbia's Pacific
          herring population.

EXT. SHORELINE — DAY

Herring are spawning in the shallow water in
a tiny bay.

                    DOUG HAY (O.S.)
          Most herring will survive three
          or four or five years, and
          probably spawn three or four
          and five times during . . .

ON A SINGLE HERRING — It struggles to survive
in a tangle of rockweed.

                    DOUG HAY (O.S.)
          . . . their lifetime before
          they're eaten.

RESUME — Hay interview.

                    DOUG HAY
          So the average mortality rate
          of a herring could be probably
          20 or 30 percent per year.

EXT. ROCKWEED SHORELINE — DAY

Herring frantically try to spawn right on the
edge and get pushed up onto the shore with
the waves.

                    DOUG HAY (O.S.)
          It varies a little bit from one
          area to the next,

A FEW HERRING — work their way along the
rockweed, laying eggs.

                    DOUG HAY (O.S.)
          . . . but the average herring
          could probably expect to live
          six or seven years.

EXT. OCEAN (DRONE) — DAY

A few commercial fishing boats are scattered
over the fishing grounds.

                    GRANT SCOTT (O.S.)
          This fish is so much more
          valuable for all of those
          industries than it is . . .

FROM THE SIDE — The herring wiggle in the net
as they are pulled toward the back of the
boat.

                    GRANT SCOTT (O.S.)
          . . . to grind them up in, into
          fish farm food and sell,

VERY TIGHT — Thousands of fish are thrashing
about in the net at the back of the boat.

Herring are caught in two ways during the sac-roe fishery: by seine or gillnets. Gillnetters place nets in the area where herring are about to spawn and the herring are caught by their gills in the nets' mesh. This is an underwater photo of herring caught in a gillnet. (Photo: Doug Hay)

A net full of herring being hauled over the back of a boat. The fish are shaken out of the net and scooped into the hold of the boat. (Photo: Scott Alpin)

Herring are also caught by seiners during the sac-roe fishery. A large
net is set around a school and slowly tightened. The herring are brailed
with smaller mesh baskets into the hold of the boat. (Photo: Doug Hay)

                    GRANT SCOTT (O.S.)
          . . . sell one small part of
          it to the Japanese market for
          sushi.

RESUME — Scott interview.

                    GRANT SCOTT
          We just don't think that's a
          higher and best use . . .

EVEN CLOSER — The seine net is full of
herring.

                    GRANT SCOTT (O.S.)
          . . . for a publicly owned,
          very valuable resource.

WIDER — on two sea lions trying to get at
the herring in the net.

EXT. ROCKY SHORELINE — DAY

Boats are bobbing in the surf, trying to
catch herring. Gulls are everywhere.

INT. UNIVERSITY OF VICTORIA CLASSROOM (GREEN
SCREEN) — DAY

Establish. Interview.

Green screen plate: Hundreds of birds,
including eagles, waiting to prey on spawning
herring.

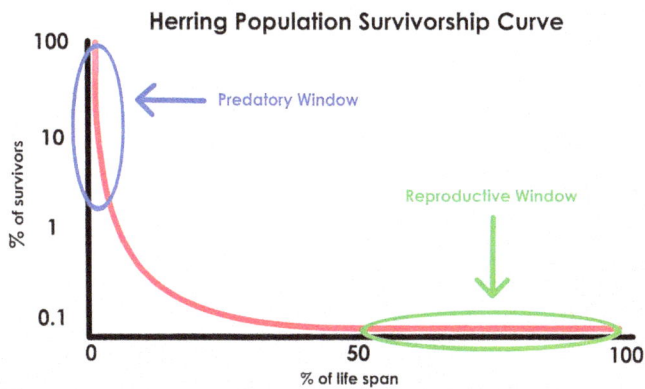

Herring Population Survivorship Curve

Prey species like herring suffer large losses in the first stages of life and only a small percentage survive to reproduce. This graph shows the herring's life history. Unfortunately, humans target the herring's reproductive window, which can have a big impact on herring's ability to restock the next generation of fish. (Graph: Pacific Coast Entertainment Ltd.)

LOWER THIRD: Dr. Thomas E. Reimchen, Biology Professor, University of Victoria.

> THOMAS E. REIMCHEN
> The reason they're producing
> 5,000 eggs or whatever it might
> be, right, is because of this
> incredible mortality,

GRAPH — titled the "Herring Population Survivorship Curve" with percent of survivors on the vertical axis and percent of life span on the horizontal axis.

The high early mortality in herring is shown by the dramatic drop in survivorship in just the first few weeks of a herring's life.

> THOMAS E. REIMCHEN (O.S.)
> . . . age-specific mortality in
> their first week, second week,
> third week a slightly different
> predator . . . fourth week,
> etc.

HERRING — flip in the shallows near shore.

A SEA LION — emerges from frothy water with a herring in its mouth.

> NARRATOR (V.O.)
> This means that most of
> the individuals in a prey
> population like herring don't
> survive to reproduce.

RESUME — Reimchen interview.

>   THOMAS E. REIMCHEN
>   That survivorship curve, that J
>   shape, right? Is a consequence
>   of predation.

RESUME SURVIVORSHIP GRAPH — The survivor
line completes and the reproductive window
and predatory window show up as highlighted
circles. The reproductive window shows on the
horizontal part of the graph near the end of
the herring's life. Then the predatory window
appears at the beginning of the herring's
life.

>   THOMAS E. REIMCHEN (O.S.)
>   The vast majority of species
>   when they hit their reproductive
>   window are outside the predatory
>   window in most instances.

A KILLER WHALE — takes a breath as it swims
past a gillnetter.

A GREAT BLUE HERON — spies a herring in the
shallows and grabs it in its bill.

>   THOMAS E. REIMCHEN (O.S.)
>   That's how they co-evolved with
>   their predators over 20 million
>   years.

RESUME SURVIVORSHIP GRAPH — The predator
window disappears and the reproductive
window morphs into the harvest window.

>   THOMAS E. REIMCHEN (O.S.)
>   So we come along and we make
>   it illegal basically to capture
>   young fish . . .

RESUME — Reimchen interview.

>   THOMAS E. REIMCHEN
>   . . . and we target
>   specifically the oldest?

EXT. CALM BAY — DAY

Fishermen on their seiner brail herring out
of their skiff and over a bunch of totes on
the deck of the larger boat.

THEN CLOSER — on the brail basket full of
herring hovering over the deck.

A BIT WIDER — The fishermen release the
herring into a tote.

>   THOMAS E. REIMCHEN (O.S.)
>   It's not really too complicated
>   why so many fish species have
>   been unable to cope with this
>   type of drawdown.

55.

TIGHT ON THE TOTE — Most of the herring are
dead, but a few continue to flip their tails.

                                    FADE OUT:

FADE IN:

EXT. PRINCE WILLIAM SOUND — DAY

Establish. A beautiful landscape surrounds
the sound. A man in a small motor boat
cruises toward a small village across the
bay.

                WOODY MORRISON (O.S.)
            I worked for environmental
            protection in Alaska . . .

RESUME — Morrison interview.

                WOODY MORRISON
            . . . and there were major sea
            runs of herring in the Bering
            Sea area and . . .

UNDERWATER — A school of herring dart in and
out of milt clouds at the edge of a spawn
event.

                WOODY MORRISON (O.S.)
            . . . herring were competitors
            for feed with the pollock.

INT. OCEAN — DAY

A school of pollock mingle on a reef.

                WOODY MORRISON (O.S.)
            Well, pollock has a higher
            dollar value than the
            herring . . .

EXT. FISHING GROUNDS — DAY

Commercial fishing boats are setting nets.

MORE BOATS — are heading out to the fishing
grounds.

                WOODY MORRISON (O.S.)
            . . . so of those seven runs,
            the Fisheries people permitted
            three of them to be fished
            until they crashed,

ONE BOAT — pulls two skiffs.

                WOODY MORRISON (O.S)
            . . . assuming this would give
            more pollock,

RESUME — Morrison interview.

                WOODY MORRISON
        . . . but it didn't.

EXT. BEACH — DAY

A dead seagull.

          WOODY MORRISON (O.S.)
    Seabirds were falling out of
    the sky. They were starving to
    death.

A FUR SEAL — is eating another seal.

          WOODY MORRISON (O.S.)
    Fur seals were turning cannibal
    because they had no food. They
    start killing and eating one
    another.

SEA LIONS — dive off a rock into the water.

          WOODY MORRISON (O.S.)
    So you can't balance one off
    for the other.

RESUME — Morrison interview.

          WOODY MORRISON
    All these things are related,
    and we need each other to
    survive is how it's said,

EXT. ROUGH OCEAN — DAY

Commercial seiners are competing with killer
whales and thousands of birds for the herring
catch.

>   WOODY MORRISON (O.S.)
>   . . . so there's no one thing
>   that is more important than
>   another.

A CLOSER ANGLE — The killer whales are
dangerously close to the boats. Eagles circle
the ships.

>   WOODY MORRISON (O.S.)
>   In Haida, there's a saying that
>   says life is like walking on
>   the edge of a knife blade.

SEAGULLS — try to grab herring off the back
of the boat. The scene is chaotic.

>   WOODY MORRISON (O.S.)
>   If you don't watch your step,
>   you can slip and fall off the
>   world.

A SEA LION — grabs a salmon from inside a
seine net.

>   WOODY MORRISON (O.S.)
>   So when we fall off that knife
>   blade, we fall into a world of
>   chaos.

SEA LIONS — swarm a cluster of seine boats
and hundreds of birds fly around them.

CLOSER — The sea lions jam between two boats.

A BUNCH OF SEA LIONS — Some are inside and others are outside a net.

FROM ABOVE — Sea lions are between the boats.

ONE SEA LION — comes up for air but stays inside the net. We hear it growl.

UNDERWATER — A sea lion is trapped in the net. It tries to swim deeper to escape, but it's stuck.

OTHER SEA LIONS — bob around the net, looking for an easy catch of herring.

A SEAGULL — plunges into the water, grabs a herring and takes off.

                                        FADE OUT:

FADE IN:

EXT. SWIFT CREEK — DAY

A small creek in the Squamish watershed dries up each year. Two Squamish Streamkeepers, Jonn Matsen and Ted Domachowski, are catching coho fry and putting them in laundry totes. The streamkeepers routinely move the coho to a place called Dave's Pond, which is a safe place for the fish to rear and move down the Cheakamus River to the ocean.

                    NARRATOR (V.O.)
          In 1993, the Department of
          Fisheries and Oceans . . .

IN THE TOTE — The fry are kept in the cool
stream water, which is well oxygenated.

                    NARRATOR (V.O.)
          . . . initiated the Salmon
          Enhancement Program to engage
          with communities to try and
          restore freshwater aquatic
          habitats.

JONN MATSEN — fishes the fry out of the tote
and places them in a pair of five-gallon
pails.

TIGHTER — Matsen and Domachowski balance out
the number of fry in each pail.

INT. OCEAN — DAY

A school of salmon swim up from the deep.

                    NARRATOR (V.O.)
          This program was in response
          to the alarming declines of
          Pacific salmon populations
          since the mid-1980s.

JONN MATSEN — carries the pails full of fry
up the bank to the road.

INT. MATSEN'S HOUSE — DAY

Establish. Interview. Dr. Jonn Matsen is a retired naturopath with a passion for the environment.

LOWER THIRD: Dr. Jonn Matsen, Herring Coordinator, Squamish Streamkeepers.

> JONN MATSEN
> I'm a streamkeeper because I was a commercial fisherman for two years and then a, uh, sports fisherman all my life really.

EXT. STEEP TRAIL — DAY

The streamkeepers carry the pails of fry to a place lower down the stream that doesn't dry up.

AS THEY PASS BY — we can see the pails are equipped with bubblers to keep the fry comfortable during transport.

> JONN MATSEN
> The salmon stocks collapsed and, uh, I decided I would like to try and help bring them back.

EXT. MAMQUAM SPAWNING CHANNELS — ANOTHER DAY

Coho fry swim in shallow water.

The streamkeepers wade through the channel.
They are using a makeshift seine net on poles
to scoop fry out of a pool of water that was
once a flowing spawning channel.

>   NARRATOR (V.O.)
>   Many people on the west
>   coast joined streamkeeper
>   organizations to help restore
>   salmon runs.

THE FRY — are poured into a pail.

TIGHTER — The fry swim around in the pail.

>   NARRATOR (V.O.)
>   From rescuing juvenile salmon
>   from drying creeks,

EXT. LITTLE STAWMUS CREEK — DAY

A beaver dam blocks the fish's passage,
especially adult salmon's passage, to the
pond above the dam.

>   NARRATOR (V.O.)
>   . . . to the notching of beaver
>   dams so adult salmon can reach
>   the spawning grounds,

ON A NOTCH — The streamkeepers have created an opening so fish can bypass the dam during high-water events.

TWO STREAMKEEPERS — join Jonn Matsen to survey a drying channel for fry.

> NARRATOR (V.O.)
> . . . streamkeepers everywhere keep busy each year, from when the fry emerge in the spring . . .

INT. STREAM — DAY

Coho fry are mingling in a spectacular mountain stream.

INT. ANOTHER STREAM — DAY

Adult coho salmon hover over a gravel spawning bed.

> NARRATOR (V.O.)
> . . . until the last salmon spawns well into the winter months.

ANOTHER SPAWNER — darts upstream.

LOOSE EGGS — lie in a calm spot in a pool. A few fry have just emerged from the eggs.

EXT. TRAIL — DAY

Jonn Matsen and another streamkeeper carry
pails of fry from a drying creek through a
patch of salmonberry bushes.

MORE STREAMKEEPERS — use beach seines to
recover fry from a small pool of water in a
drying creek bed.

> NARRATOR (V.O.)
> In spite of everyone's effort,
> after nearly 30 years of stream
> work . . .

ALEVINS — by the thousands are in hatching
trays in a hatchery.

> NARRATOR (V.O.)
> . . . and hatcheries pumping
> out millions of fry,

INT. HIGH FALL SPAWNING CHANNEL — DAY

A majestic pair of coho spawners drift
through a deep pool. They share the space
with a sea-run cutthroat trout.

> NARRATOR (V.O.)
> . . . salmon populations have
> not improved and continue their
> decline towards extinction.

RESUME MATSEN — and Domachowski as they
unload pails of fry from a vehicle.

IN A CREEK — One of the men pours the coho fry into a safe stream so they can continue rearing before they migrate to the ocean.

> NARRATOR (V.O.)
> The streamkeepers began to wonder if salmon populations were declining because of a lack of food in the ocean.

FADE TO:

EXT. SQUAMISH ESTUARY (DRONE) — DAY

The estuary sits in front of magnificent mountains, including one known as The Stawamus Chief.

> JONN MATSEN (O.S.)
> Herring used to come right into the Squamish Harbour, massive numbers.

UNDERWATER — A school of herring cruise slowly by.

INT. SQUAMISH STUDIO (GREEN SCREEN) — DAY

Establish. Interview.

Green screen plate: The Mamquam Blind Channel near the Squamish Harbour boat launch.

LOWER THIRD: John Buchanan, Conservationist, Squamish, BC.

> JOHN BUCHANAN
> You know, I have to go back to
> my childhood again, because as
> a child . . .

UNDERWATER — A herring jig bumps up and down.

MOMENTS LATER — a herring is hooked and lifted out of the water.

> JOHN BUCHANAN (O.S.)
> . . . I used to go down to the
> small boat harbour here and jig
> for herring.

PHOTO — of the Squamish public wharf in the mid-1960s. Locals are jigging herring between the boats.

> JOHN BUCHANAN (O.S.)
> We would come home with a
> bucket of herring and, uh, Mom
> would fry it up like bacon and,
> and that's just what we did
> when we were kids, right?

PHOTO — of Eric and Glen Andersen sitting on a bluff overlooking the Squamish Estuary with the lumber mill in full operation. Circa 1964.

PHOTO — from the mouth of the Mamquam Blind Channel. It's obvious that logging is a major activity in this waterway.

> JONN MATSEN (O.S.)
> Squamish decided that they didn't want to be a one-season logging town any more. They wanted industry to come in.

PHOTO — from across the Mamquam Blind Channel with the mill and beehive burner.

> JONN MATSEN (O.S.)
> A big sawmill was built right next to the herring spawning ground.

HEADLINE — "Weldwood fined for fish kill," *Squamish Times*, March 14, 1974.

> JONN MATSEN (O.S.)
> Someone moved a lever the wrong way and . . .

RESUME — Matsen interview.

> JONN MATSEN
> . . . wood preservatives flushed out into the spawning channel and killed 500,000 herring.

PHOTO — looking down the Mamquam Blind Channel with the mill on the east shore and barges being filled with sawdust.

> JONN MATSEN (O.S.)
> Basically that was the
> beginning of the end.

PHOTO — of the chemical plant known as Nexen. The plant is butted up against the shoreline on the southern tip of the Mamquam Blind Channel.

INT. MAMQUAM BLIND CHANNEL — DAY

It's a dirty environment filled with cables, industrial waste and creosote pilings covered in muck. There's no sign of life here. We now realize that the strange and mysterious environment in the opening of the film is the Squamish Estuary.

> JOHN BUCHANAN (O.S.)
> Between around 1966 to 1971
> they lost 40 tons of mercury
> into the surrounding areas.

PHOTO — of the chemical plant.

> JOHN BUCHANAN (O.S.)
> And that comes from the
> chemical plant that operated
> here in Squamish.

HEADLINE — "BCR port could be deep sixed,"
Ahead of the Times, September 15, 1992.

PUSHING IN — on the words "The mercury could
be so hot that it can't be dealt with."

> JOHN BUCHANAN (O.S.)
> Apparently these herring that
> we were fishing as kids had a
> high mercury content in them.

UNDERWATER — The muck and pilings in the
channel show an environment in deep trouble.

MOVING UP — A tire is stuck in the pilings
and there is just a small shaft of light
coming through the mess above.

> JOHN BUCHANAN (O.S.)
> So at some point in the '70s
> they just disappeared.

RESUME — Buchanan interview.

> JOHN BUCHANAN
> Around about the same time, our
> coastline started to get very
> heavily industrialized.

EXT CATTERMOLE SLOUGH — DAY

Old pilings are clustered along the shoreline
at the mouth of the Cattermole Slough. The
Squamish Terminals shipping dock fills the
landscape.

The Squamish Streamkeepers prepare to lower a boat off the Squamish Terminals west dock. Their plan on this day was to cover the concrete pilings with weed control cloth wraps. (Photo: Scott Renyard)

THROUGH OLD PILINGS — A giant ship sits at the Squamish Terminals.

REVERSING — The slough is lined with muck and sparse sea grasses. The grasses blow in the winter wind.

> JONN MATSEN
> For 30 years there were no herring in upper Howe Sound in the Squamish area.

THE STAWAMUS CHIEF BLUFF — is covered in snow. It is still winter.

PANNING DOWN — onto a patch of dense sea grasses blowing in the wind.

FURTHER DOWN — white dots are sprinkled on the grass.

> JONN MATSEN (O.S.)
> We started looking around, and sure enough we'd see little patches of grass with eggs on them floating around.

ANOTHER ANGLE — The eggs are sparse, but they are definitely eggs.

VERY TIGHT — The dots are herring eggs.

                    NARRATOR (V.O.)
          The discovery of these eggs,
          as few as they were, created a
          buzz of excitement among the
          streamkeepers.

ANOTHER PATCH — of grass with a few more
eggs. They are white, fresh and alive.

                    NARRATOR (V.O.)
          This meant that some local
          herring had survived,

UNDERWATER — A small school of herring swim
slowly through the estuary.

                    NARRATOR (V.O.)
          . . . and there was hope
          something could be done to
          bring them back to their former
          levels of abundance.

PHOTO — of the lumber mill on the Mamquam
Blind Channel.

The same view enjoyed by the Andersen
brothers in the 1960s. But this time the mill
is modern. No beehive burner. And stacks of
fresh lumber.

PHOTO — of the mill, now modernized. It
reflects the commitment to using the Mamquam
Blind Channel for industry.

Anglers fish for herring off the Squamish Harbour dock in 1960. Herring started to disappear from the Squamish Estuary in the 1970s as industrial development ramped up, mostly in the Mamquam Blind Channel. (Source: Squamish Library Archive)

In 1942, the Squamish Estuary was mostly in pristine condition with just the old rail terminal and its replacement. Both of these wharves were mothballed and a new terminal was constructed. The new terminal is known as the Squamish Terminals. (Photo: Province of British Columbia)

By 1968, the estuary, especially the Mamquam Blind Channel, had been dredged and was an industrial hub for logging and a chemical plant. (Source: BC Archives)

          ERIC ANDERSEN (O.S.)
      By the mid-'80s a solution was
      to divide the estuary in half.

INT. SQUAMISH STUDIO (GREEN SCREEN) — DAY

Establish. Interview.

Green screen plate: The western half of
the estuary, which is clearly not used by
industry.

**LOWER THIRD: Eric Andersen, Historian,
Squamish, BC.**

          ERIC ANDERSEN
      Easterly half to be used
      for development, industry,
      water-dependent industry and
      transportation,

EXT. SQUAMISH ESTUARY WEST — DAY

Establish the natural condition of this side
of the estuary.

          ERIC ANDERSEN (O.S.)
      . . . and the west side for
      conservation.

A GOOSE — leads 14 goslings along the beach.

ZIPPING ALONG — the entrance to the Mamquam Blind Channel. Its shoreline is a collection of pilings, collapsed wharfs, riprap and mostly badly degraded habitat.

> ERIC ANDERSEN (O.S.)
> Many of the people who are actively interested in the estuary and its use and its protection,

AN OLD PILING WALL — lines part of the Mamquam Blind Channel shoreline.

> ERIC ANDERSEN (O.S.)
> . . . and the estuary management plan for that matter, will say,

CRUISING BY — a wall of steel pilings.

> ERIC ANDERSEN (O.S.)
> . . . well, things have changed. We don't have BC Rail anymore,

PHOTO — of the sawmill with Mount Garibaldi in the background.

> ERIC ANDERSEN (O.S.)
> . . . sawmill's gone, industry's gone.

INT. MAMQUAM BLIND CHANNEL — DAY

A FEW HERRING — have found a small patch of
rockweed and are spawning.

> ERIC ANDERSEN
> Now we see that the herring
> like to spawn in the cleaner
> water of the east side.

EXT. MAMQUAM BLIND CHANNEL (DRONE) — LATER

Looking north from the mouth of the channel.
We see a large body of water lined with
logging activities on both sides.

> NARRATOR (V.O.)
> The east side of the estuary is
> now known as the Mamquam Blind
> Channel. But at one time it
> was actually the mouth of the
> Mamquam River.

EXT. HOWE SOUND (1929) — DAY

A large storm whipping up whitecap waves
howls up the sound toward Squamish.

THE WIND — is so strong the tops of the waves
are blown into a mist.

> EDITH TOBE (O.S.)
> In the 1920s, there was a large
> storm event.

This aerial photograph shows that during the spring freshet, the Mamquam Channel water is cleaner than the silty water from the Squamish River. However, herring will spawn throughout the estuary when the Squamish River water is not as silty, prior to the spring freshet. The concern has been that herring are more attracted to the cleaner water of the Mamquam Blind Channel, where most of the industrial activity occurs. Recently, most of the industrial activities have been shuttered and replaced with housing. These changes should clean up the channel waters, which would benefit herring. (Photo: Province of BC)

Prior to the flood of 1929, the Mamquam River flowed into what is now known as the Mamquam Blind Channel. (Source: Province of British Columbia; map annotated by Pacific Coast Entertainment Ltd.)

After the flood, the Mamquam River was diverted directly into the Squamish River. The Mamquam River Reunion project involved punching a hole in the dike to add more water to Logger's Lane Creek to improve the water conditions in the Mamquam Blind Channel and rewater the old river path. (Source: Province of British Columbia; map annotated by Pacific Coast Entertainment Ltd.)

PHOTO — Aerial photo modified and animated
to show the old path of the Mamquam River as
it morphs into its current path. (Source: BC
Government air photo, 1949.)

> EDITH TOBE (O.S.)
> During that large storm, the
> Mamquam River, which originally
> flowed down what is currently
> the Mamquam Blind Channel . . .

PHOTO — Aerial view of the Squamish area and
the original path of the Mamquam River before
the high water event. Then the original path
disappears and the new path appears on the
photo.

> EDITH TOBE (O.S.)
> . . . was redirected along
> its current channel into the
> Squamish River.

EXT. MAMQUAM BLIND CHANNEL (GREEN SCREEN) —
DAY

Establish. Interview. Edith Tobe is a
professional biologist who lives in Squamish,
British Columbia.

Green screen plate: A section of the Mamquam
River flows along the dike that blocked the
river from following its pre-1929 path.

LOWER THIRD: Edith Tobe, Executive Director, Squamish Watershed Society.

                    EDITH TOBE
          The people living in Squamish
          at that time diked the river
          off to prevent it from going
          back down to the Mamquam Blind
          Channel.

EXT. STAGNANT POND — DAY

Groundwater pools in the old swale, with
just a trickle of water making its way to the
Mamquam Blind Channel.

                    EDITH TOBE (O.S.)
          So the only water going back to
          the Mamquam Blind Channel since
          the 1920 event was basically
          groundwater.

ON STAGNANT WATER — Leaves float on the
shallow water.

TIGHTER — The water seems perfectly still.

                    EDITH TOBE (O.S.)
          In 2005, the Watershed Society
          working . . .

RESUME — Tobe interview.

> EDITH TOBE
> . . . with Fisheries and Oceans
> Canada again approached the
> District of Squamish in a
> partnership to put a hole in
> the dike,

EXT. MAMQUAM REUNION ENTRANCE — DAY

A pipe with a control valve allows Mamquam River water to enter the top end of a new channel to rewater the Mamquam Blind Channel.

> EDITH TOBE (O.S.)
> . . . diverting a small portion
> of the flows back into the
> Logger's Lane's system and back
> into the Mamquam Blind Channel.

ON THE WATER — The new stream flows over the gravel.

SKUNK CABBAGES — line the new creek further down.

INT. MAMQUAM BLIND CHANNEL — DAY

Establish. There is nothing at the bottom but the remains of an old shopping cart.

> NARRATOR (V.O.)
> But the channel was in shocking
> condition.

A SAILBOAT — has sunk and is mostly underwater.

ANOTHER BOAT — rests at the bottom of the channel.

RESUME — circling the oil spill boom to find . . .

                    NARRATOR (V.O.)
          Sunken boats and hazardous
          debris littered the sea floor.

DISGUSTING OIL SLUDGE — surrounds an abandoned floathouse.

                    NARRATOR (V.O.)
          And frequent oil slicks hugged
          the shoreline where herring
          could spawn.

A FEW HERRING — swim past the wharf.

FOUR STREAMKEEPERS — in a Zodiac head across the channel.

TRAVELLING — along the shoreline with logs, oil sludge and abandoned boats. It's a mess.

                    NARRATOR (V.O.)
          So the streamkeepers rallied
          their supporters and undertook
          a major clean-up. They hoped
          they could make the channel
          less toxic to herring eggs.

The Mamquam Blind Channel was lined with old wharves, sunken and abandoned boats, logging debris and lots of creosote pilings. It was very hazardous for herring eggs. (Photo: Scott Renyard)

Shorelines altered like this one provide very little natural vegetation or plants.
Note that there is just a small amount of bladder wrack kelp (*Fucus vesiculosus*) on the
one rock that is a preferred surface for herring spawners. (Photo: Scott Renyard)

Another example of the industrial waste and debris that covered
much of the shoreline of the Mamquam Blind Channel, making it difficult
for herring to find safe spawning surfaces. (Photo: Scott Renyard)

Herring spawners resorted to spawning on a dead tree's roots
in the Mamquam Blind Channel because there were no other
clean surfaces on which to lay their eggs. (Photo: Scott Renyard)

This battery was found at low tide in the Mamquam Blind Channel. It is just
one example of the trash and debris found at the bottom of the waterway
that could be detrimental to herring and their eggs. (Photo: Scott Renyard)

THE STREAMKEEPERS — push the remains of a speedboat off the east bank of the channel.

PASSING BY — a rusty old tug surrounded by an oil containment boom.

A HALF-SUNKEN SAILBOAT — juts out of the water.

THE HULL OF A SPORT BOAT — is dragged up the boat launch. It's upside down.

OIL SLUDGE — swirls on the surface of the water like an abstract art painting.

A TUG — yanks a yellow boat off the shore. It then drags it across the channel to the boat ramp.

WHIPPING — past an orange oil containment boom. Pilings hold it in place. The irony of oily toxic pilings somehow helping the situation seems expected.

ANOTHER ANGLE — More abandoned and derelict boats along the shore and even more oily sludge covering the water's surface.

THE REFLECTION OF A HULL — is obscured by the oil sludge floating on the water.

RESUME TOWING — The yellow boat is dragged up the ramp.

ON THE SAILBOAT — It's now at the ramp.

A BACKHOE — is crushing the sailboat up.

THE BACKHOE — is now smashing up another boat and loading it into a Dumpster. Lots of garbage falls out onto the ground.

FINALLY — a chunk of a boat is shoved into the Dumpster.

                                        FADE OUT:

FADE IN:

INT. OCEAN — DAY

The murky water looks like it is home to submerged trees. Why would trees be growing underwater?

CLOSER — There are branches and needles. Yes, they are trees. In fact, they are hemlock trees.

                    WOODY MORRISON (O.S.)
               When we were getting the eggs,
               we'd cut a small tree that's
               maybe . . .

LOOKING DEEPER — The trees look like they've been caked in something slimy.

                    WOODY MORRISON (O.S.)
               . . . about six feet tall.

TIGHTER MONTAGE — The coating is fresh white herring eggs, thousands of them.

> WOODY MORRISON (O.S.)
> And they would spawn on it.

RESUME — Morrison interview.

> WOODY MORRISON
> Then we'd pull the whole tree
> up and cut the branches loose
> from the trunk.

SUDDENLY — one of the egg-covered trees begins to shake. It is pulled up out of frame.

HERRING EGGS — are caked on hemlock branches.

VERY CLOSE — on the eggs sprinkled on the hemlock branches.

> NARRATOR (V.O.)
> First Nations use hemlock to
> gather herring eggs for food.

THE EGGS — are white, clean and alive.

> NARRATOR (V.O.)
> But the streamkeepers had
> another idea.

EXT. MAMQUAM BLIND CHANNEL (DRONE) — DAY

Flying quickly just above the water heading to the mouth of the channel.

PHOTO — of Jonn Matsen and Cal Harknell under the power line easement near Squamish.

SUPERSCRIPT: February 2006.

           NARRATOR (V.O.)
      They got permission from the
      local power company . . .

PHOTO — of Matsen and Harknell wrestling with a small hemlock they just cut down with a chainsaw.

           NARRATOR (V.O.)
      . . . to harvest a few small
      hemlock trees that had been
      slated for removal.

PHOTO — of Harknell cutting a tree with his chainsaw while another streamkeeper, Patrick MacNamara, watches him.

PHOTO — of Matsen, Hartnell, MacNamara and others prepping the hemlock trees next to the Squamish Harbour boat launch.

           NARRATOR (V.O.)
      They worried that their channel
      clean-up would not be enough to
      help the herring.

PHOTO — of MacNamara placing the butt of a
hemlock tree in a cinder block and attaching
it with rope.

PHOTO — of Jonn Matsen, Cal Hartnell and Eric
Andersen at another log boom on the west side
of the Mamquam Blind Channel, just south of
the boat launch. They are unloading a couple
of hemlock saplings with cinder blocks tied
to the stump ends.

                    NARRATOR (V.O.)
          Their new plan was to submerge
          hemlock from wharves and log
          booms to add safe spawning
          surfaces in the channel.

PHOTO — of Matsen, Hartnell and Andersen.
They have taken some of the trees by small
skiff to log booms on the east side of the
Mamquam Blind Channel. Matsen is poised to
drop a tree into the water.

PHOTO — The whole group are now on a floating
dock. A stack of trees are ready to be
deployed into the water.

EXT. MAMQUAM BLIND CHANNEL — MOMENTS LATER

A tree with a block on it is tossed into the
water. It sinks slowly out of sight.

INT. MAMQUAM BLIND CHANNEL — MOMENTS LATER

The tree sinks deeper into the depths.

> NARRATOR (V.O.)
> Unfortunately, the
> streamkeepers had overlooked a
> vital detail. For the hemlock
> plan to work, herring needed
> to be actively spawning in the
> area.

ANOTHER TREE — sinks deeper into the water.

                                        FADE TO:

PHOTO — of Jonn Matsen holding a rope that is
presumably attached to one of the hemlocks.
Brandon Hartnell is looking on.

> NARRATOR (V.O.)
> So when the streamkeepers went
> back,

INT. MAMQUAM BLIND CHANNEL (UNDERWATER) —
MOMENTS LATER

A silhouette of what seems like a lone tree
shrouded in the green water of the channel
looks almost magical and mysterious.

SLIGHTLY WIDER — It now looks like a tree,
but barely. Something is wrong.

> NARRATOR (V.O.)
> . . . they discovered the trees
> were not covered in eggs like
> they had hoped.

Natural rocky shorelines provide plenty of bladder wrack kelp (*Fucus vesiculosus*), also known as rockweed, which is safe and suitable for herring eggs to be laid on. (Photo: Scott Renyard)

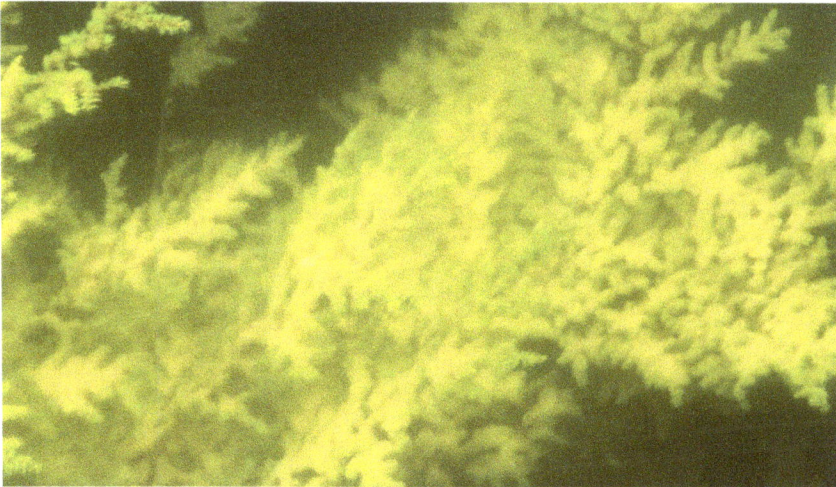

First Nations fishers place different marine plants or hemlock boughs in the water to harvest herring eggs. This photo shows a hemlock tree underwater and covered in herring eggs. (Photo: Scott Renyard)

This feather boa kelp, *Egregia menziesii*, is often used by First Nations fishers to gather herring eggs. (Photo: Pacific Coast Entertainment Ltd.)

Herring will often spawn so thick on hemlock and other plants placed by First Nations fishers that the plant is barely discernible. (Photo: Pacific Coast Entertainment Ltd.)

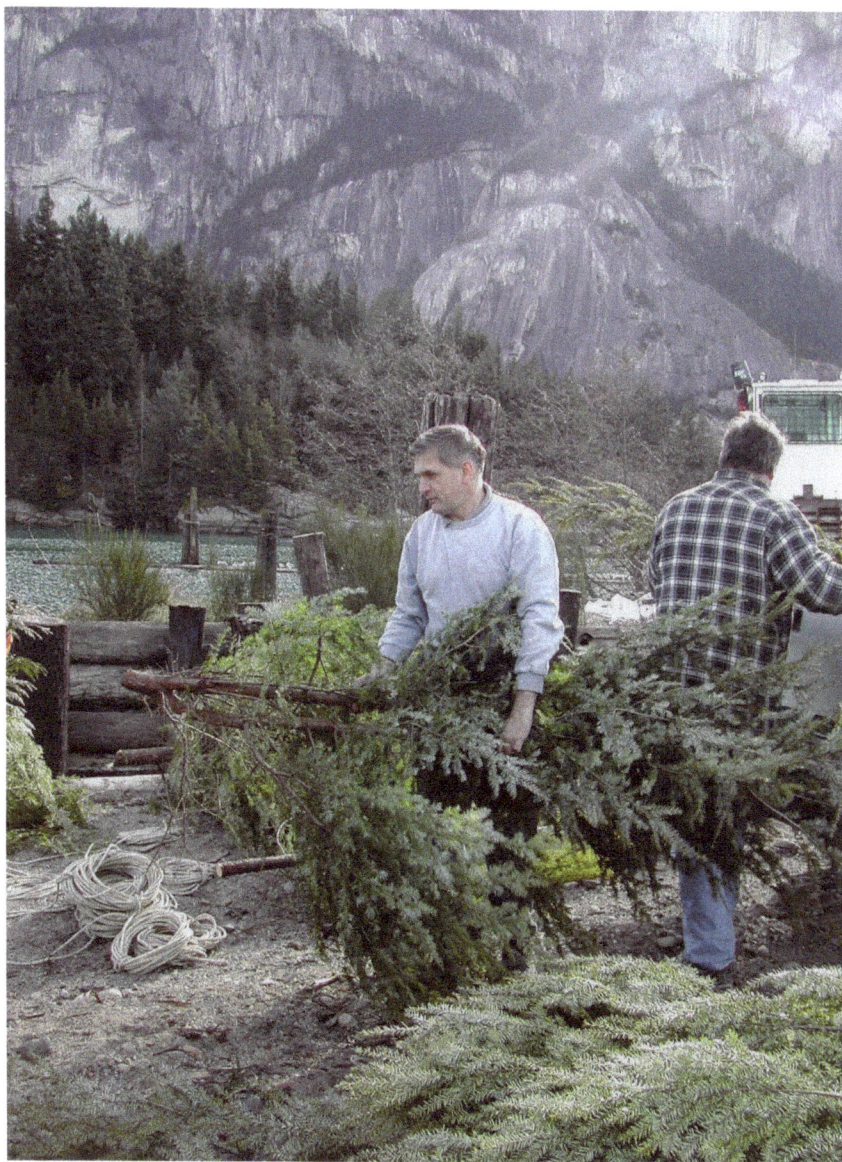

Jonn Matsen and Cal Hartnell offload hemlock saplings they harvested from under a local power line. These hemlocks were being removed to keep the power line clear and were available for the streamkeepers' herring project. (Photo: Scott Renyard)

CLOSER — The tree is covered in brown slime.

                    NARRATOR (V.O.)
          They were covered in slime and
          their experiment was a bust.

TWO MORE TREES — are covered in the slime.

PHOTO — from the top of the Stawamus Chief,
with the whole estuary in view. The Squamish
Terminals dominate the centre of the estuary.

EXT. SQUAMISH TERMINALS — DAY

Establish.

EXT. SQUAMISH ESTUARY SHORELINE — DAY

Looking across the slough to the Squamish
Terminals at the modest guard shack.

                    NARRATOR (V.O.)
          A few weeks later, Jonn Matsen
          got a call from an old friend
          who worked at the Squamish
          Terminals.

NEXT TO THE SHORE — a few herring are
spawning in the shallows.

                    NARRATOR (V.O.)
          He told Matsen he could see
          herring spawning in the water
          near the guard shack.

UNDERWATER — A bed of rockweed sways with the movement of the waves. There are no herring in sight.

                    NARRATOR (V.O.)
          By the time the streamkeepers
          arrived, the herring were gone.

PHOTO — of Jonn Matsen crouching down and taking a photo of eggs on riprap near the guard shack.

                    NARRATOR (V.O.)
          But not all was lost. After
          looking around, they found
          fresh eggs on the rockweed
          along the shoreline.

TIGHT — on some rockweed. The algae is sprinkled with herring eggs.

                    NARRATOR (V.O.)
          The herring were there all
          right.

EXT. SQUAMISH TERMINALS EAST DOCK — DAY

A strange-looking floating dock is located at the north end of the Squamish Terminals east dock. It has four large openings with raised edges. They are rearing pens that the Department of Fisheries and Oceans was using at that time to rear chinook smolts as part of yet another program designed to help declining wild salmon populations.

                    NARRATOR (V.O.)
          They also discovered a floating
          wharf with salmon-rearing
          pens at the north end of the
          Terminals east dock.

A DFO ZODIAC — passes by the pens.

                    NARRATOR (V.O.)
          The Department of Fisheries
          and Oceans were using the
          pens . . .

PHOTO — of Jack Cooley lowering some supplies
with a rope. The salmon-rearing pens are in
clear view.

                    NARRATOR (V.O)
          . . . to raise salmon smolts in
          salt water to try and increase
          their rate of survival.

SPLASH — A hemlock tree hits the water.

                    NARRATOR (V.O.)
          They then got permission to
          submerge hemlock trees from the
          pen wharf to see if the herring
          were looking for more surfaces
          to spawn on.

EXT. SQUAMISH TERMINALS EAST DOCK — DAY

The streamkeepers gather on the deck of the
east dock. Then,

The Squamish Streamkeepers attached ropes and cinder blocks to the hemlock trees and submerged them in the Mamquam Blind Channel. The goal was to add safe substrates for herring to spawn on. The experiment was unsuccessful because herring spawners were not active in the area. The branches of the hemlocks became covered in barnacles and algae instead of herring eggs. (Photo: Scott Renyard)

A guard at the Squamish Terminals could see herring spawning along this shoreline. The streamkeepers went to investigate and found herring eggs on the riprap close to the guard shack (seen on the left of the photo). This discovery pushed the streamkeepers to move their project closer to the Squamish Terminals. (Photo: Scott Renyard)

Herring spawned on the bladder wrack kelp (*Fucus vesiculosus*) on the Squamish Terminals spit near the guard shack. They also found herring eggs on sea grasses at the mouth of the Cattermole Slough. This confirmed that herring were still in the area, and the streamkeepers ramped up their efforts to try to help the local herring population. (Photo: Scott Renyard)

The Squamish Streamkeepers placed hemlock saplings at this DFO salmon pen wharf to see if herring were looking for something clean to spawn on. Note the green and yellow colours on the pilings under the fixed wharf. (Photo: Scott Renyard)

WIDER — They then march in single file along
the edge of the dock.

>   NARRATOR (V.O.)
>   One day, after working on
>   their hemlock experiment, they
>   decided to go under the dock to
>   look for herring. And they were
>   surprised by what they found.

RESUME — Matsen interview.

>   JONN MATSEN
>   And we saw this yellow goop on
>   the creosote pilings.

EXT. SQUAMISH TERMINALS EAST DOCK — DUSK

The streamkeepers make their way down the
riprap embankment and disappear under the
dock with flashlights in hand.

UNDER THE DOCK — The beam of a flashlight
bobs its way toward camera.

LOOKING TOWARD THE WATER — Flashlight beams
bounce off the creosote pilings. Many of
them are not black, but tan or yellow.

>   NARRATOR (V.O.)
>   The streamkeepers discovered
>   a spooky environment filled
>   with the aroma of chemicals
>   emanating from more than 800
>   creosote pilings.

PHOTO — of a piling covered in yellow goop. Patrick MacNamara reaches out and touches the goop.

> NARRATOR (V.O.)
> And the goop was on dozens of the pilings.

PHOTO — of another piling. It's not quite as yellow. The goop looks more like little pellets.

PHOTO — of the pellets closer up. They look like eggs.

> JONN MATSEN (O.S.)
> And just as a hunch, we thought that might be dead herring eggs.

PHOTO — of one piling on which the eggs are barely pellets and look like goop.

PHOTO — of one piling covered in tan-coloured eggs. The eggs are dead and mouldy.

PHOTO — of eggs up close. They are brown and dead.

> NARRATOR (V.O.)
> They were herring eggs, and tens of millions of them had died.

While standing on the DFO salmon-rearing pens at low tide, the streamkeepers noticed some of the pilings were yellow. They decided to go under the docks and discovered that herring were spawning on the pilings. In this photo, streamkeeper Patrick MacNamara puts his hand on a piling where the eggs died and turned into yellow goop. (Photo: Scott Renyard)

The streamkeepers later discovered that the yellow goop turns into a rusty pink mould. Many of the pilings were covered in the pink mould. The streamkeepers began to use the mould as a way to locate pilings that herring were using as a spawning substrate. (Photo: Scott Renyard)

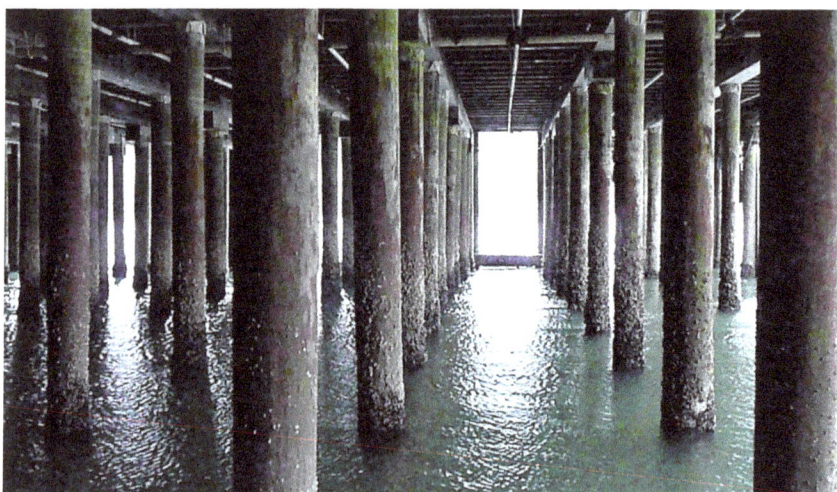

A wider shot of pilings at the old Squamish Terminals east dock with rusty pink mould on them, indicating herring had used most of them in the past for spawning. (Photo: Scott Renyard)

Herring also spawn on timber retaining walls supported by pilings if they are submerged at high tide. The pink mould in this case looks purple and is a sign of a spawn event in which the eggs died. (Photo: Scott Renyard)

INT. SQUAMISH TERMINALS EAST DOCK — LATER

The rows of pilings are clearer now. And many of them have a rusty pink mould on them.

TIGHTER — Nine pilings in a tight row have varying degrees of mould stains on them.

> NARRATOR (V.O.)
> They also realized that the pilings with the rusty pink mould were evidence of past spawning events gone bad.

LOOKING DOWN — the full length of a row of pilings. The closest ones have lots of rusty pink patches on them.

LOOKING UP — One piling has mould covering about two metres of its height.

> FADE TO:

PHOTO — of a man squeezing herring eggs from a ripe female into a neat row on an aluminum plate in a laboratory.

PHOTO — of another ripe female having her eggs squeezed onto a line of creosote.

> DOUG HAY (O.S.)
> We've done tests by incubating herring eggs on creosote in the lab,

RESUME — Hay interview.

> DOUG HAY
> . . . and eggs that touch the
> creosote by and large don't
> survive. They die during
> development.

INT. OCEAN — DAY

A school of herring have found a piling and
are spawning on it.

THE HERRING — delicately spawn over mussels
that have grown on this piling.

> DOUG HAY (O.S.)
> We do know that herring will
> spawn sometimes on other things
> in the water.

LOOKING UP — at the bottom of a boat hull
covered with eggs.

> DOUG HAY (O.S.)
> Floating objects will sometimes
> attract them.

ANOTHER PILING — This one is covered in
barnacles. The herring don't seem to care and
are happily spawning on them.

> DOUG HAY (O.S.)
> Pilings will attract them.

LOOKING DOWN — the piling. The herring swim up while depositing eggs.

EXT. MAMQUAM BLIND CHANNEL — DAY

The channel is anything but natural. Pilings, riprap and collapsed docks dominate the waterway.

> NARRATOR (V.O.)
> This discovery changed the direction of the streamkeepers' herring project from improving spawning habitat in a compromised estuary . . .

TIGHT ON EGGS — that are clearly dying.

> NARRATOR (V.O.)
> . . . to a project focused on protecting herring eggs from creosote pilings.

VERY TIGHT — Some of the eggs have collapsed, but some look alive.

INT. OCEAN — DAY

The pilings look like tree trunks in a foggy, green forest.

> NARRATOR (V.O.)
> But why do the herring spawn on the pilings in the first place?

INT. OCEAN — DAY

A beautiful bull kelp forest looks from this
angle a little like the pilings and wharf
overhead.

> NARRATOR (V.O.)
> Could it be that pilings look
> like a kelp forest to herring?

INT. OCEAN — DAY

From underneath another wharf, we see the
structure has a similar look to the kelp
forest.

HERRING — swim randomly around a single
piling.

> NARRATOR (V.O.)
> Or perhaps our man-made
> structures simply replace the
> plant cover herring need so
> they can avoid being preyed
> upon.

EXT. SHORELINE — DAY

An eagle swoops down out of the sky. It
snatches a herring out of the water and flies
off with it in its mouth.

UNDERWATER — A herring school swim in a tight
ball to avoid predators. A bird picks one
herring off and swims away with it.

THE SCHOOL — dart away, swimming for their lives.

> NARRATOR (V.O.)
> Herring, like all forage fish, are on the run their entire lives. Each phase of their life happens quickly.

ON A ROCKY OUTCROP — Herring are plastering the aquatic plants with herring eggs.

> NARRATOR (V.O.)
> And it starts with the eggs.

TIGHT — on rockweed covered with fresh eggs.

EXT. BAY (DRONE) — DAY

The mildly milted water is still translucent, and a herring school can be seen swimming in it.

> NARRATOR (V.O.)
> Herring often spawn in bays and inlets during slack tides . . .

IN THE SHALLOWS — herring spawn right at the edge of the shoreline in thick rockweed.

> NARRATOR (V.O.)
> . . . to ensure the male's milt will linger long enough to fertilize the female's eggs.

The Squamish Streamkeepers devised a plan to wrap weed control cloth around the creosote pilings to try to protect herring eggs from coming into contact with the toxic characteristics of creosote. To get access to the pilings, they had to time their wrapping project to coincide with the lowest tides. (Photo: Scott Renyard)

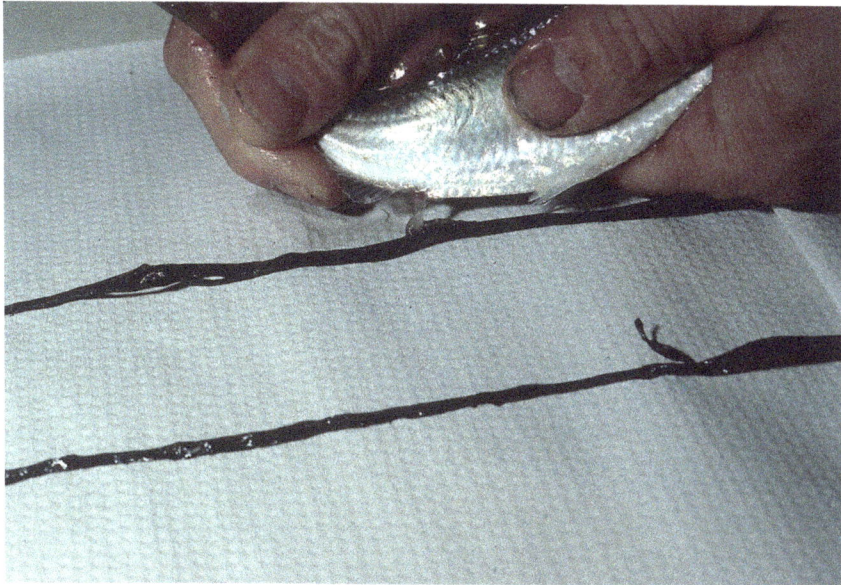

The DFO performed laboratory experiments by placing herring eggs on creosote. These experiments confirmed that creosote is toxic to eggs. (Photo: Doug Hay)

INT. OCEAN — DAY

The herring have plastered their eggs on a
piling.

CLOSER — The mass of eggs covers not only the
surface of the piling but also the barnacles.

> NARRATOR (V.O.)
> Once fertilized, the eggs
> undergo cell divisions within
> hours.

ONE SECTION — is all white, fresh eggs.

TIGHTER — on the translucent eggs. The
beginning of an embryo can be seen inside
them.

> NARRATOR (V.O.)
> And in just three days, the
> embryo encircles the yolk.

AN EMBRYO — jerks and spins inside an egg.
And then another one moves inside its egg.

> NARRATOR (V.O.)
> By day 6, the embryo is well
> formed and it begins to move
> inside the egg.

INT. OCEAN — DAY

A large boulder is covered in rockweed. The
rockweed is covered in fresh eggs.

CLOSE UP — Many of the eggs are opaque white.
They look like they have died.

> NARRATOR (V.O.)
> But even on natural substrates,
> egg mortality is often 65
> percent and can be as high as
> 80 percent.

EXT. BEACH — DAY

Hundreds, if not thousands, of birds are
feasting on herring eggs.

> NARRATOR (V.O.)
> Then predators like birds,
> which can eat 40 percent or
> more of the spawn at a given
> location, prevent millions of
> eggs from reaching the next
> stage of the herring life
> cycle.

INT. OCEAN — LATER

A piling is home to barnacles and lots of
herring eggs.

SOME OF THE EGGS — have survived, and
hatchlings are emerging from them.

ANOTHER SPOT — The eggs are eyed. A hatchling
squirms free from its shell.

>   NARRATOR (V.O.)
>   The remaining eggs, depending
>   on temperature, hatch into
>   larvae between 10 and 21 days.

NEARBY — One herring larva tries desperately
to free itself from its egg shell.

INT. OCEAN — NIGHT

A beam of light shining into the darkness
reveals a tremendous hatch of herring larvae.

CLOSER — The light shines off the eyes of the
newly born larvae like two headlights on a
wriggling string of vermicelli.

>   NARRATOR (V.O.)
>   Remarkably, the eggs from a
>   spawn event typically hatch
>   within a few hours of each
>   other, and the emerging larvae
>   form a wriggling mass that is
>   the first significant step
>   towards creating a new school.

EVEN CLOSER — Dozens of eyes seem to stare
back at us.

INT. OCEAN — DAY

At an eel grass bed, jellyfish float around.
It's the perfect spot for them to find
herring larvae to feed upon.

AT NIGHT — The water is filled with arrow
worms hunting for prey.

> NARRATOR (V.O.)
> Herring larvae are easily
> preyed upon by a whole host of
> animals, including jellyfish
> and arrow worms.

EXT. ROCKY SHORELINE — DAY

A wave crashes onto the rocks.

> NARRATOR (V.O.)
> And even with all the early
> losses from predation,

UNDERWATER — Algae jerk violently in the
current.

> NARRATOR (V.O.)
> . . . herring larvae can
> suffer even greater losses if
> they are swept away in tidal
> currents . . .

HERRING LARVAE — are caught up in a violent
current and are tossed around.

> NARRATOR (V.O.)
> . . . and lose contact with
> preferred inshore food.

INT. OCEAN — LATER

A school of young juvenile herring swim
quickly, darting in different directions.
They are so young they are still translucent.
Suddenly they are spooked and dart away.

                    NARRATOR (V.O.)
          This can result in mortality at
          an alarming 99.5 percent.

HERRING JUVENILES — seem abundant, with
thousands swimming together. But many of them
will be lost to predation before they reach
maturity.

INT. NEAR SHORE OCEAN — DAY

A school of herring look like a time-lapse
cloud as they dart away from a number of
large fish predators.

                    NARRATOR (V.O.)
          And as they become schools,
          they are pursued for the rest
          of their lives by salmon and
          marine mammals higher up the
          food chain.

UNDERWATER — Herring scatter as a humpback
whale emerges from the dark blue water with
its mouth open wide, ready to gobble them up.

                                   FADE OUT:

FADE IN:

EXT. MAMQUAM BLIND CHANNEL — DAY

Travelling quickly across the water. A log
dump fills the far shoreline.

                    NARRATOR (V.O.)
          Since early natural mortality
          can be so high,

INT. MAMQUAM BLIND CHANNEL — MOMENTS LATER

A few herring are spawning on a piling.

                    NARRATOR (V.O.)
          . . . human activities that
          impact spawning areas are an
          extra insult to a . . .

FROM THE SIDE — Herring shimmy up and down
the piling, laying eggs.

THE MURKY — green water makes it all seem
surreal.

                    NARRATOR (V.O.)
          . . . species that plays such
          a crucial role in the marine
          ecosystem.

INT. MAMQUAM BLIND CHANNEL (RE-ENACTMENT) — A
LONG TIME AGO

There is a nice bed of eel grass.

In a good year, herring spawn can cover approximately 500 kilometres of BC coastline. The milt from the males changes the colour of the water to a brilliant turquoise during a spawn event. (Photo: Jeremy Mathieu)

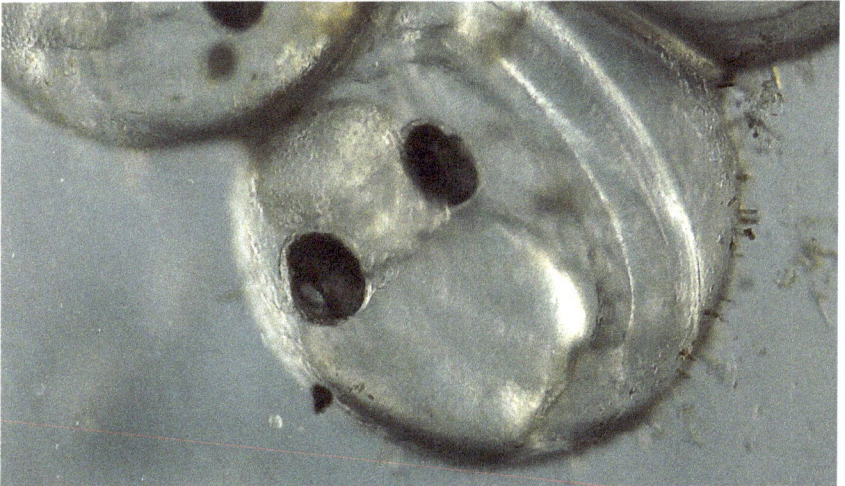

The herring gestation period is a short one. Once an egg is laid, the embryo undergoes cell divisions within hours. This egg is about 10 days old and the embryo is fully developed and has eyes. (Photo: Scott Renyard)

DISSOLVE TO:

TODAY — the eel grass bed is full of brown algae and barely alive.

> NARRATOR (V.O.)
> In many estuaries, eel grass
> that was once plentiful is now
> virtually non-existent.

NEAR THE LOG DUMP — bark and debris cover the ocean floor and have choked out any semblance of a vibrant ecosystem.

A TIGHTER VIEW — shows just how bad it is.

> NARRATOR (V.O.)
> Around logging operations,
> debris and bark littered across
> the bottom choke out underwater
> vegetation.

DEEPER WATER — The bottom is smooth with a brown matt-like covering that is more like a slimy brown carpet than a vibrant sandy bottom full of life.

> NARRATOR (V.O.)
> And to accommodate boat
> traffic, dredging removes or
> buries . . .

A PAIR OF CRABS — hide between brown riprap rocks.

                    NARRATOR (V.O.)
          . . . nearly all of the animals
          and plants that normally live
          on the bottom.

INT. SQUAMISH TERMINALS OCEAN — LATER

The shade from the massive structure makes
the bottom under the wharf dark. It's not
much brighter than a moonlit night, even
with the sun trying to peek under the dock.

ELSEWHERE — it's so dark the pilings are mere
silhouettes in dark green water.

                    NARRATOR (V.O.)
          And shadows cast from log
          booms, wharves and other
          shoreline structures block
          sunlight and prevent the growth
          of marine plants.

AT THE BOTTOM — Except for a few Dungeness
crabs and the odd sea anemone attached to
a rusty pipe, the ocean floor is nearly
lifeless.

A COUPLE OF CRABS — stir up the silt on the
ocean floor as they scoot away.

                    NARRATOR (V.O.)
          All of these impacts can turn
          an estuary into a lifeless
          moonscape unable to support a
          vibrant marine ecosystem.

INT. OCEAN — DAY

Herring are spawning on a mussel-encrusted piling.

A FEW HERRING — weave around each other, trying to get closer to the piling.

> DOUG HAY (O.S.)
> Herring avoid silt in the water. It could have the effect of dissuading herring from spawning there in the first place.

RESUME — Hay interview.

> DOUG HAY
> Or if the eggs are there and silt comes after the fact,

FOUR UNDERWATER PILINGS — have crud on their surface. Many of the eggs are dead.

THE DEAD EGGS — are covered with silt.

> DOUG HAY (O.S.)
> . . . then it might have a deleterious effect on the eggs themselves.

VERY CLOSE — Many of the eggs are opaque and dead.

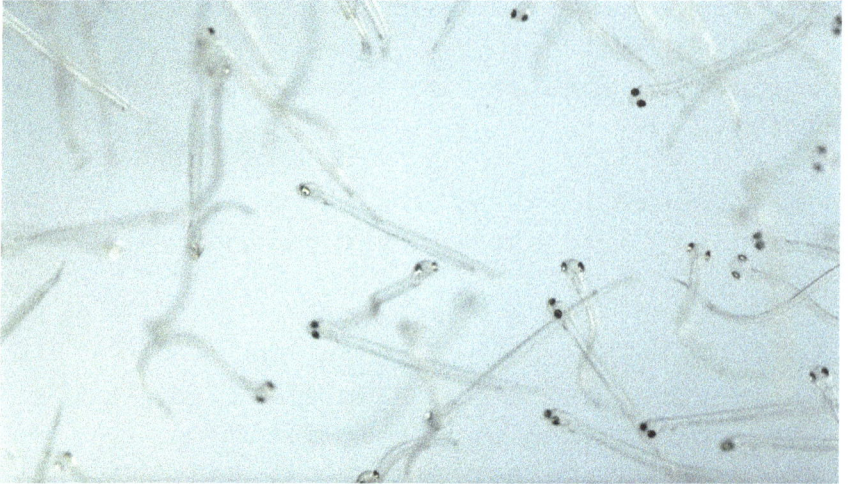

The eggs hatch within 10–21 days, depending on the water and air temperature. A herring hatchling is known as a larva and does not resemble an adult herring at all. It remains translucent, like a piece of vermicelli with eyes, for the first few months of its life. (Photo: Pacific Coast Entertainment Ltd.)

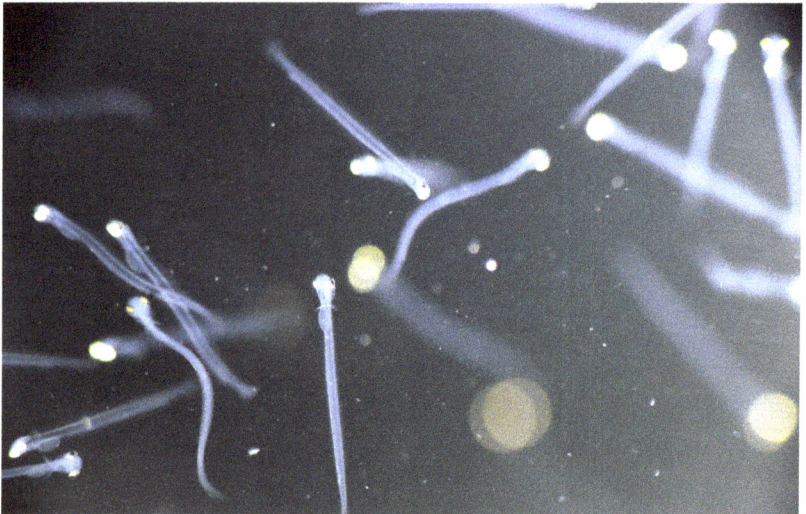

These translucent herring larvae have two large eyes and respond eagerly to light. If they are exposed to light at night, they will rise to the surface of the water. This behaviour stays with them throughout their lives and can lead them into trouble. Fishermen used to shine large lights at night, a practice known as pit-lamping, to draw herring into their nets. Pit-lamping was so effective it was banned in the 1960s. Despite this, open net pen fish farms are permitted to light up their farms at night. Herring are attracted into the pens, where they are eaten. (Photo: Scott Renyard)

EXT. ROCKWEED SHORELINE — DAY

The water is thick with milt. A few herring
cruise overhead in the shallows.

>   NARRATOR (V.O.)
>   What we do know is that herring
>   eagerly spawn on aquatic
>   vegetation . . .

INT. ANOTHER BAY — LATER

The marine plants are sprinkled with fresh
herring eggs.

>   NARRATOR (V.O.)
>   . . . and rocks that appear
>   clean.

VERY CLOSE — on a bed of rockweed. The
herring eggs on these plants are alive and
healthy. They move back and forth with the
wave action.

ON A BED — of an algae called japweed
(*Sargassum muticum*). It is thick with eggs.

>   NARRATOR (V.O.)
>   And most of these surfaces
>   are found in the intertidal
>   zone that under natural
>   conditions . . .

A SINGLE FROND — of japweed waves back and
forth with the wave action.

                    NARRATOR (V.O.)
          . . . is subject to wave action
          as the tides move in and out
          each day.

INT. OCEAN — DAY

The herring are spawning on a piling that is
partially covered with barnacles and mussels.

                    NARRATOR (V.O.)
          So when spawners encounter
          pilings, and they are the only
          silt-free surfaces available,

FROM THE SIDE — of the piling. The herring
continue to rub their bellies on it.

ON ONE HERRING — It curves its tail away
from the surface of the piling to press its
eggs on the surface.

                    NARRATOR (V.O.)
          . . . herring spawn on them,
          often with tragic results.

INT. MAMQUAM BLIND CHANNEL — DAY

The piling is covered with eggs. But they are
brown. Something is wrong.

EXT. MAMQUAM BLIND CHANNEL (DRONE) — DAY

Flying north toward the Squamish Harbour public dock. The Mamquam Blind Channel is clearly an ecologically compromised estuary.

SUPERSCRIPT: October 2010.

>   NARRATOR (V.O.)
>   In October 2010, the Canadian
>   Government announced that they
>   would . . .

ABOVE THE PUBLIC DOCK — It has three main floats.

>   NARRATOR (V.O.)
>   . . . rebuild and expand the
>   Squamish Harbour dock located
>   in the Mamquam Blind Channel.

PANNING UP — a group of new creosote pilings.

>   NARRATOR (V.O.)
>   The design called for the use
>   of new creosote pilings.

INT. SQUAMISH DOCK, OCEAN — DAY

Panning up from the ocean floor, we see some pilings are mostly devoid of living organisms except for a few live barnacles.

Jack Cooley visits the Squamish Harbour dock when it was undergoing a renovation and expansion. New floats are attached to the existing dock and ready to be installed once new pilings are put in place. (Photo: Scott Renyard)

New creosote pilings were installed for the Squamish Harbour dock. Their black, clean-looking surfaces, free of silt and debris, are attractive to herring spawners, especially in areas where natural substrates are sparse or compromised in degraded environments. The streamkeepers warned officials that this expansion might attract herring spawners and kill their eggs. (Photo: Scott Renyard)

> NARRATOR (V.O.)
> The streamkeepers issued a
> public warning that the new
> creosote pilings would attract
> herring spawners and it could
> be bad news. But, in spite
> of the warning, the project
> proceeded as planned.

FADE TO:

SUPERSCRIPT: March 12, 2011.

LATER — Herring are spawning heavily on the new pilings.

> NARRATOR (V.O.)
> The following spring, over 10
> million eggs were laid on each
> of the new pilings.

AFTER THE SPAWN — the normally black pilings are now white. They are covered with eggs.

PANNING ACROSS — The herring eggs are not white, but tan in colour.

> NARRATOR (V.O.)
> After one week, the eggs began
> to change colour.

WIDER — Four pilings are covered in brown eggs.

TIGHTER — The eggs are now tan and no longer translucent. They are opaque.

> NARRATOR (V.O.)
> After two weeks, the eggs
> went brown with the stain of
> creosote.

PULLING BACK — The eggs on the pilings have become a mustard-coloured mess. They are dead.

> NARRATOR (V.O.)
> After three weeks, the eggs
> began to rot and became a gooey
> slime.

TIGHT — on one piling. The eggs have dried out. None have survived.

> NARRATOR (V.O.)
> All the eggs died.

EXT. SQUAMISH HARBOUR DOCK — ANOTHER DAY

A crew is working to wrap a set of pilings.

ON A PAIR OF HANDS — tightening a zap strap.

> NARRATOR (V.O.)
> Efforts were made to design and
> build wraps to block any more
> herring from spawning on the
> pilings.

Just as the streamkeepers predicted, the herring spawned heavily on the new creosote pilings. The eggs absorb the creosote and, when the tide changes and the eggs are exposed to the air, they dry out and die. (Photo: Scott Renyard)

Approximately 10 million herring eggs were spawned on 15 new creosote pilings (February 2011). One hundred and fifty million eggs perished in just one spawn event. It's clear that new creosote pilings are harmful to herring eggs and should not be placed in areas used by herring spawners. (Photo: Scott Renyard)

WIDER — The group has created an awkward wrap
made of pool noodles, zap straps and cloth.

UNDERWATER — looking up. The larger wrap
covers the pilings, but the design seems
fragile and anything but durable.

BACK TO — the dead eggs on the pilings.

> NARRATOR (V.O.)
> But the damage was done: 150
> million eggs died from that
> single event. It was a disaster
> that could have been avoided.

PUSHING IN — on the goopy mess.

ACROSS — the piling. The goop is dark brown
in places.

FADE OUT:

FADE IN:

EXT. SQUAMISH TERMINALS (DRONE) — DAY

Moving across the sky above the Terminals,
first the west dock and then the east dock
come into view.

RESUME — Matsen interview.

**LOWER THIRD: Dr. Jonn Matsen, Herring
Coordinator, Squamish Streamkeepers.**

JONN MATSEN
We had permission from the
Squamish Terminals to do some
experimenting on trying to
bring these eggs to fruition.

JONN MATSEN — at the east dock with Lyle Wood
and Hugh Kerr. He is giving the others some
instructions.

NARRATOR (V.O.)
The streamkeepers decided to
wrap the pilings with some kind
of material . . .

MATSEN — pulls supplies out of the back of
his truck.

NARRATOR (V.O.)
. . . that would block the
creosote from coming into
contact with the eggs.

Then Matsen puts on his waders.

JONN MATSEN
I'll be under the dock and I'll
yell directions.

He guides the forklift over to the side of
the dock.

JONN MATSEN
That's the right stuff there.

                    FORKLIFT OPERATOR (O.S.)
                They must've got it out for you
                already.

MOMENTS LATER — Brad Ray and Jonn Matsen
head under the dock.

                    NARRATOR (V.O.)
                So on a cold January day in
                2008, the streamkeepers set
                to work under the Squamish
                Terminals east dock . . .

INT. SQUAMISH TERMINALS — LATER

The streamkeepers enter the area under the
dock. It's like stepping into a cave.

                    NARRATOR (V.O.)
                . . . to see if this would work.

PANNING ACROSS — the dark space. Pilings
block most of the light.

JONN MATSEN — is on a ladder, wrapping one
piling with black plastic.

A SECOND PILING — is wrapped in screen door
material.

TWO STREAMKEEPERS — wrap a piling with weed
control cloth.

The Squamish Streamkeepers needed to wrap the pilings in January of each year so the wraps would be fresh and in place just before the spawning season began in late January. Unfortunately, the lowest tides were after dark, so they geared up like miners with headlamps and flashlights to do their work. (Photo: Scott Renyard)

Initially, the streamkeepers only wrapped the pilings that they could reach from shore. (Photo: Scott Renyard)

                    NARRATOR (V.O.)
          The materials chosen were black
          plastic, screen door mesh and
          weed control cloth.

BRAD RAY — whacks a piling with a stapler to
secure the wrap to it.

LATER — Matsen leads the group out from under
the dock.

RESUME — Matsen interview.

                    JONN MATSEN
          And lo and behold, the, uh,
          next time we went under the
          dock they were covered with
          herring eggs . . .

INT. TERMINALS EAST DOCK OCEAN — DAY

The wrapped piling is caked in fresh herring
eggs.

                    JONN MATSEN (O.S.)
          . . . and that was the
          beginning.

ON THE EGGS — It's obvious that the herring
covered every centimetre of the piling in
several layers of eggs.

VERY CLOSE — All the eggs are alive.

The Squamish Terminals east dock had 11 rows of creosote pilings that were beyond the reach of the streamkeepers from the shore at low tide. They decided to use small boats to reach some of these pilings so they could wrap them as well. (Photo: Scott Renyard)

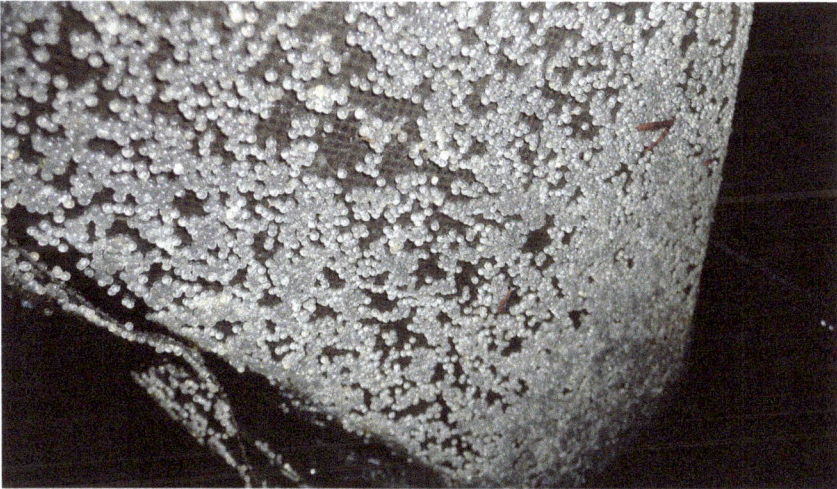

The streamkeepers' first wrapping project included only a few pilings as an experiment. After the next spawn event, it was clear the herring preferred wrapped pilings to unwrapped pilings. (Photo: Scott Renyard)

Jonn Matsen takes photographs of fresh eggs on a wrapped piling. Most of the eggs hatched out from that first spawn, and the success motivated the streamkeepers to expand their project and wrap more pilings. (Photo: Scott Renyard)

The streamkeepers experimented with several materials but settled on weed control cloth when they noticed the herring preferred it to black plastic, black felt and screen door mesh. They even stapled the material into sheets to try to expand the safe surface area for herring eggs. This experiment failed because wave action tore the cloth and it sank to the ocean floor. (Photo: Scott Renyard)

The streamkeepers redesigned the rigid sheets into a float line, which looked a lot like a commercial fishing net. In this case, the netting was replaced with weed control cloth. The herring spawned on the material, much to the delight of Jonn Matsen, the group's coordinator. (Photo: Scott Renyard)

The herring liked the material so much that several layers of eggs covered the float line material. This posed another problem for the streamkeepers: the egg density was too thick, and they were worried some eggs would suffocate. They also discovered that a thick coating of eggs on a float line would peel off with wave action and likely suffocate in the silt-covered ocean floor when the eggs fell to the bottom. (Photo: Jonn Matsen)

WIDER — One of the pilings is being spawned on. It's an amazing dance, but it looks almost alien.

LOOKING DOWN — the piling. Herring dart up and down and across the surface as they deposit their eggs.

HERRING SPAWNERS — vie for space as they swim toward camera.

                    NARRATOR (V.O.)
          Not only did the herring spawn
          heavily on the wraps, it was
          clear they preferred them over
          the unwrapped surfaces.

PHOTO — of Jonn Matsen taking a photo of the herring eggs on one of the wrapped pilings.

PHOTO — of fresh herring eggs.

                    NARRATOR (V.O.)
          So when most of the eggs
          hatched, the experiment was
          declared a huge success.

PHOTO — of herring eggs closer up.

PHOTO — of more herring eggs on a wrapped piling.

INT. SQUAMISH TERMINALS EAST DOCK — THE NEXT YEAR

Jonn Matsen climbs a ladder to wrap a piling.

> JONN MATSEN
> Every year from there on we
> kept adding more and more and
> more.

ON HIS HAMMER — as he taps in nails to secure the wrap.

WIDER — Lyle Wood, Cal Hartnell and Jonn Matsen have finished adding wraps and are working their way to the north end of the east dock.

> NARRATOR (V.O.)
> However, the streamkeepers
> discovered something strange
> after the first spawn on the
> wrapped pilings.

LOOKING AT — a piling wrapped in black plastic. The herring have avoided spawning on it.

ON ANOTHER PILING — a wrap made of screen door material has few eggs on it.

> NARRATOR (V.O.)
> The herring barely spawned on
> the screen door material and
> completely avoided the shiny
> black plastic.

INT. OCEAN — DAY

A school of herring cruise along the ocean
floor.

> DOUG HAY (O.S.)
> In terms of the surface areas
> that herring spawn on,

RESUME — Hay interview.

> DOUG HAY
> . . . they seem to avoid
> vegetation that has slime on
> it,

THE OCEAN FLOOR — A patch of light green
algae looks shiny and slippery.

> DOUG HAY (O.S.)
> . . . and that seems to make
> sense because the eggs are
> easily removed.

NEARBY — A pile of loose eggs are at the
mercy of the water's movement.

INT. SQUAMISH TERMINALS EAST DOCK — ANOTHER DAY

The streamkeepers return to add more weed control cloth.

> NARRATOR (V.O.)
> So they decided to drop the shiny materials and just use the weed control cloth.

THE WORK — continues under the dock.

> FADE TO:

Jonn Matsen passes behind a piling with a badly torn wrap.

> NARRATOR (V.O.)
> The next year, they discovered even more problems.

CLIMBING OUT — from under the dock. The streamkeepers retreat from assessing the damage to their weed control wraps.

> NARRATOR (V.O.)
> They found that the weed control cloth had torn away because of waves and driftwood banging around the pilings.

PANNING DOWN — a wrapped piling. The wrap is stained and appears to be covered in oil.

                    NARRATOR (V.O.)
      And surprisingly, creosote had
      seeped through the wraps and
      killed most of the eggs laid on
      the contaminated cloth.

EXT. SQUAMISH TERMINALS EAST DOCK — ANOTHER
DAY

The streamkeepers enter from the north end of
the east dock with new materials.

                    NARRATOR (V.O.)
      The streamkeepers realized they
      needed to use a more durable
      material and found a harder
      plastic that seemed to work.

ON THE PILINGS — They are now caked with eggs
from top to bottom.

THE EGGS — on the pilings are several layers
thick.

                    NARRATOR (V.O.)
      The herring loved the new
      material, but maybe too much.
      Now the egg density was getting
      dangerously thick.

CLOSER AND FROM THE SIDE — The layers of eggs
are impressive.

PANNING UP — one piling, we can still see the
stainless steel fasteners, although barely.

                    DOUG HAY (O.S.)
          There's laboratory and field
          studies that indicate that at
          very high densities there's a
          danger of suffocation of eggs,
          particularly the ones that
          tend to be at the bottom of the
          pile.

ANOTHER PILING — has a sparse amount of eggs
on it. Sticklebacks are busy eating the eggs.

                    DOUG HAY (O.S.)
          There's a trade-off in herring
          eggs between having very low
          densities . . .

RESUME — Hay interview.

                    DOUG HAY
          . . . where the predators can
          take most of your eggs and very
          high densities where you might
          lose eggs from . . .

INT. SQUAMISH TERMINALS OCEAN (UNDERWATER) —
LATER

Three pilings are covered in eggs, but they
appear to be discoloured and not doing well.

                    DOUG HAY (O.S.)
          . . . mutual interference
          through suffocation and low
          oxygen.

TIGHT ON THE EGGS — Most of them are dead.

EXT. SQUAMISH TERMINALS EAST DOCK — DAY

An aluminum skiff is lowered from the deck down to the water.

A RAMP — is lowered and the streamkeepers and a team of volunteers make their way down to the salmon pen floating dock.

BOATS — are loaded with wrap materials and the crew pair up in the skiffs. They are planning to wrap the pilings that they couldn't wrap from shore.

                    NARRATOR (V.O.)
          The streamkeepers ramped up
          their efforts and wrapped as
          many pilings as possible to
          encourage the spawners to
          spread their eggs on more
          pilings.

THE CREW — paddle their way under the dock.

A VOLUNTEER — secures the wrap around a piling with a rope.

BRAD RAY — pulls a rope around another wrap.

ERIC ANDERSEN — grips the piling and hammers a wrap into place.

LOOKING BETWEEN THE PILINGS — Three boat
crews are covering a row of pilings with the
new material.

                                    FADE TO:

INT. SQUAMISH TERMINALS OCEAN — DAYS LATER

Herring are spawning on one of the newly
wrapped pilings.

                    NARRATOR (V.O.)
            But the trouble was, there
            was no way to know how many
            spawners were coming.

THE FRANTIC — spawners cover the piling
quickly. It's chaotic.

                    NARRATOR (V.O.)
            And if another wave of spawners
            came before the first eggs
            hatch, what happens then?

A LONE PILING — is partially covered with
unhatched eggs.

EXT. PILING — LATER

After the tide drops, some older eggs are
exposed to the air. Their thick coating and
colour indicate that they are not fresh eggs.

ON NEW EGGS — They look fresh and are not laid on top of other eggs.

> DOUG HAY (O.S.)
> Remarkably, they often don't spawn over old herring eggs.

RESUME — Hay interview.

> DOUG HAY
> So it sometimes happens that you have waves of spawning in an area.

HERRING — are spawning on another piling. It's surreal and magical to watch, but likely tragic.

FROM ANOTHER ANGLE — The herring are spawning on a fresh surface and not on old herring eggs.

CLOSER — It's clear all the eggs are fresh.

> DOUG HAY (O.S.)
> And you have one batch of spawning area in one week and a week later you have a little bit of a later spawn.

FROM THE WHARF — The shimmering sides of the herring give away the spawners and the eggs they are depositing on the piling.

The streamkeepers later discovered that the rusty pink mould was on every piling and beam of the Squamish Terminals west dock. These pilings, however, were made of concrete and not creosote. The streamkeepers initially thought that concrete was also toxic to herring eggs. (Photo: Scott Renyard)

Strangely, the rocks under the Squamish Terminals west dock were also covered in the rusty pink mould. This finding was a bit of a head-scratcher at the time. Why would herring eggs perish on a natural surface like rock? Was it because the water around the Terminals was contaminated? The streamkeepers later learned from John Buchanan's observation that herring eggs seem to die more easily on hard surfaces. (Photo: Scott Renyard)

                    DOUG HAY (O.S.)
          It seems that when they do that,
          the more recently deposited
          eggs are not on the older eggs.

EXT. SQUAMISH TERMINALS WEST DOCK — DAY

The concrete pilings under the west dock all
have rusty pink mould on them.

                    NARRATOR (V.O.)
          The streamkeepers noticed that
          the telltale pink mould was
          also covering the concrete
          pilings on the Terminals west
          dock.

AT THE NORTH END — The pilings are stained,
but less obviously than other pilings we've
seen. The concrete beams under the wharf
ceiling have pink mould on them.

LOOKING — between the rows of pilings. The
mould is everywhere. The pilings are stained
pink with some green algae patches as well.

CLOSER — on the crossbeams at the top. They
are all covered in pink mould.

                    NARRATOR (V.O.)
          They feared that there
          was something toxic in the
          concrete.

AND YET — another angle showing the mould.

During the spring of 2009, most of the herring eggs laid on the Squamish Terminals west dock's concrete pilings died. This spurred the streamkeepers to wrap these pilings as well. (Photo: Scott Renyard)

EXT. SQUAMISH TERMINALS OCEAN — A YEAR LATER

Establish. A view of the west dock from across the water.

> NARRATOR (V.O.)
> The following year, they
> discovered millions of eggs on
> the concrete pilings had died.

A BOAT — is carried by the streamkeepers down the riprap to the water.

JONN MATSEN — is pointing at the area to be wrapped.

> NARRATOR (V.O.)
> This tragedy pushed the
> streamkeepers to expand the
> scope of their efforts to
> include these pilings as well.

THREE STREAMKEEPERS — launch aluminum skiffs into the water.

MORE BOATS — are filled with material. They spread out to allow the streamkeepers to wrap pilings.

ONE WRAP — is pulled around a piling and attached. But how?

                    NARRATOR (V.O.)
          The problem facing them this
          time was how to attach the
          wraps to pilings without
          drilling or damaging the
          concrete.

TWO-PERSON TEAMS — use ropes to hold the
boats close to a piling. One team member uses
a rope around the piling to pull it close.
The other puts the wrap in place and secures
it.

ONE TEAM — Jack Cooley and Tom Renyard, can
reach a piling from shore. They tug a wrap
around the piling.

ANOTHER TEAM — struggles to tighten a wrap
close to the water level. Then, the wrap is
pulled apart. It's obvious now that they are
using Velcro.

                    NARRATOR (V.O.)
          After some head-scratching, the
          plan was to use Velcro strips
          and zap straps to attach the
          material to the pilings.

BACK TO — the Cooley and Renyard team. They
secure the Velcro strip.

COOLEY — runs his fingers down the Velcro
strip to fasten it.

WIDER — They've wrapped the two rows of pilings closest to shore.

INT. SQUAMISH TERMINALS WEST DOCK — THREE WEEKS LATER

The herring have found the wrapped pilings and are plastering them with eggs.

>   NARRATOR (V.O.)
>   And just like on the east
>   dock, the herring showed up and
>   plastered their eggs on the new
>   wraps.

AFTER THE SPAWN — the wraps are completely covered in eggs.

EXT. HOWE SOUND — DAY

The wind whips the ocean into whitecaps. They crash into the estuary.

>   NARRATOR (V.O.)
>   A few days later, freezing
>   winds ripped through the
>   Squamish Estuary.

INT. SQUAMISH TERMINAL EAST DOCK — A FEW DAYS LATER

The water around the pilings is calm now. But the eggs on the closest piling are yellow.

CLOSER — Three pilings have yellow-brown
eggs. Things are not looking good.

> NARRATOR (V.O.)
> The sub-zero temperatures froze
> all of the eggs that were
> exposed to air at low tide.

UNDER THE DOCK — Jonn Matsen and Jack Cooley
have their cameras out. They take pictures of
the dead eggs.

TIGHT ON THE DEAD EGGS — The eggs are going
brown, just as if they were being killed by
creosote.

> NARRATOR (V.O.)
> The streamkeepers suddenly
> realized that pilings not
> only posed a toxic threat to
> herring eggs, but eggs laid on
> them can die from extreme air
> temperatures as well.

EXT. FOULGER CREEK — DAY

The water rushes through some boulders.
Rockweed in the foreground is caked in fresh
herring eggs.

> JOHN BUCHANAN (O.S.)
> This gets flushed,

JOHN BUCHANAN — a Squamish resident and
conservationist, enters frame and lifts up
some of the kelp with eggs with his fingers.
He points to the eggs.

> JOHN BUCHANAN
> . . . there's lots of water
> going around this, uh, lots
> of oxygen. So you've got, you
> know, really good healthy eggs.

TIGHT ON FINGERS — Buchanan is holding viable
eggs on his fingers.

> JOHN BUCHANAN (O.S.)
> This is an excellent example
> where you've got heavy spawn on
> the bladder wrack. That looks
> very viable.

ON THE BARE ROCK — The eggs on the rock are
turning white and are dying or dead.

> JOHN BUCHANAN (O.S.)
> And directly behind it, it's
> dying because of lack of oxygen
> because it's so thick.

BACK TO BUCHANAN'S FINGERS — He holds up the
eggs on the kelp that are alive.

> JOHN BUCHANAN (O.S.)
> You see, this, it gets water
> on both sides. Right, it gets
> oxygenated.

On a trip with John Buchanan, April 2015, to Foulger Creek on the west side of Howe Sound, I filmed eggs from a recent spawn event. (Photo: John Buchanan)

The eggs were concentrated around the mouth of Foulger Creek on the rocks and bladder wrack kelp (*Fucus vesiculosus*). Surveys conducted by John Buchanan over several years revealed that herring spawning in Howe Sound stretches from the Mamquam Blind Channel along the west side of Howe Sound to the Defence Islands. Foulger Creek, however, experiences the heaviest spawn along this stretch every year and should be preserved as an important ecological site for the Howe Sound herring. (Photo: John Buchanan)

John Buchanan made a key observation at the Foulger Creek location. He found that herring eggs on the kelp were healthy and viable. But often eggs deposited on the rocks died. This finding made the streamkeepers realize that herring eggs on any hard surface were at risk from exposure to extreme temperatures or drying out. (Photo: Scott Renyard)

Herring eggs on the bladder wrack kelp (*Fucus vesiculosus*) at Foulger Creek were mostly viable and healthy. (Photo: Scott Renyard)

INT. SQUAMISH TERMINALS EAST DOCK (FLASHBACK) — DAY

The herring are spawning heavily on the pilings.

A PILING (FLASHBACK) — is plastered with eggs. The density looks similar to that of eggs on the Foulger Creek rocks. They look dead.

> JOHN BUCHANAN (O.S.)
> Like when you wrap pilings and too many eggs come in on the wrap because there's too much concentration in one spot.

RESUME FOULGER CREEK — Buchanan has a camera in his hand, ready to take photos.

HE LIFTS — a bladder wrack kelp plant out of the way. We can see the dead eggs on the rock behind it.

A LARGE ROCK — is covered in the rockweed. The eggs on the plants are alive.

> JOHN BUCHANAN (O.S.)
> There's no difference between what's happening here in nature and what we've been doing on the pilings.

VIABLE EGGS — are hanging off the bladder wrack kelp.

> JOHN BUCHANAN (O.S.)
> I think that there's no
> difference between aged
> concrete and rock.

ANOTHER SPOT — These eggs, seen closer up, are very much alive.

> JOHN BUCHANAN (O.S.)
> I think they react the same to
> herring spawn. And this is what
> I want to document today.

ON BUCHANAN'S HAND — He takes a photo of the dead eggs.

EXT. HOWE SOUND ROCK WALL — DAY

Travelling along the wall. There is a layer of rockweed high above the low tide. It's clear that herring like to spawn in that layer, but the eggs could easily be exposed to the air.

> NARRATOR (V.O.)
> Suddenly, protecting herring
> eggs just got a whole lot
> more complicated for the
> streamkeepers.

EXT. SQUAMISH TERMINALS EAST DOCK — DAY

The streamkeepers have stretched out a long piece of weed control cloth on the deck of the east dock.

UNDERWATER — A diver swims under the dock.

> NARRATOR (V.O.)
> Not only were they trying to
> solve problems associated with
> the pilings,

THE DIVER — pulls the rope attached to a
large piece of cloth and wraps the cloth
around a concrete piling.

THE ROPE — is tied to the piling.

> NARRATOR (V.O.)
> . . . but now they needed to
> navigate nature itself.

UNDERWATER — The cloth is pulled between the
rows of pilings. What are the divers doing?

HERRING — swim over a bed of rockweed.

> FADE OUT:

FADE IN:

EXT. ROCKY SHORELINE — DAY

Giant waves crash against the shore.

A ROUGH OCEAN — churns against the rocks. A
storm is underway.

                    DOUG HAY (O.S.)
           We often have storms in this
           part of the world,

EXT. ANOTHER ROCKY BEACH — DAY

A huge wave crashes on the shoreline.

EXT. BEACH — DAY

The storm is over. There is a row of herring
eggs on the beach. It's 2 metres wide and at
least 0.5 metres deep.

                    DOUG HAY (O.S.)
           . . . and those storms will
           dislodge herring eggs.

PANNING ACROSS — the shoreline. We see
hundreds of millions of eggs at the water's
edge.

                    DOUG HAY (O.S.)
           And sometimes we can see,
           uh, what we call windrows
           of herring eggs. Sometimes a
           couple feet deep in areas . . .

WINDROW EGGS — are piled up and mixed in with
all the bits of sticks and seaweed.

                    DOUG HAY (O.S.)
           . . . when the eggs are
           dislodged and washed up.

Herring eggs can be dislodged during storms and pushed up on the shore. This phenomenon is commonly referred to as windrow herring eggs, the billions of herring eggs that form a strip along a beach. These strips are often several metres wide and up to a metre deep. Often the eggs are washed back into the ocean, and many presumably survive. (Photo: Doug Hay)

Vertical hard surfaces without marine plants to retain moisture during low tides make herring eggs more vulnerable to hot or cold air temperatures when they try to spawn along developed shorelines. (Photo: Scott Renyard)

CLOSE ON THE EGGS — As they come into focus, we can see that most of them look alive.

                    DOUG HAY (O.S.)
          So, that would not seem to be
          an optimal place for herring to
          put those eggs.

PANNING UP — another part of the beach. Many herring eggs are stranded on the shore. There are hundreds of gulls flying just above the surf.

                    DOUG HAY (O.S.)
          Although, even when we see
          that, often those eggs are
          still alive.

RESUME — Hay interview.

                    DOUG HAY
          It's worthwhile just to leave
          those eggs there. Some of them
          will be flushed back out into
          the sea . . .

EXT. BEACH — DUSK

Rays of light filter through the clouds, lighting up the ocean under a semi-cloudy sky.

                    DOUG HAY (O.S.)
          . . . and apparently do ok.

ON EGGS — loose on the bottom of the ocean. They're being tossed around with the wave action and most of them seem to be alive.

MORE EGGS — slosh back and forth at the bottom of the ocean.

> NARRATOR (V.O.)
> Realizing that herring eggs
> should remain underwater at
> all times to avoid contact with
> extreme air temperatures,

EXT. SQUAMISH TERMINALS EAST DOCK — DAY

Jonn Matsen and Jack Cooley are pulling a long piece of weed control cloth across the dock.

> NARRATOR (V.O.)
> . . . the streamkeepers looked
> to commercial fishing for the
> next idea.

AT THE EDGE — of the dock. Matsen and Cooley feed the weed control cloth over the side.

FROM THE SIDE — The new creation, now called a float line, is fed over the side and under the dock.

> NARRATOR (V.O.)
> They created a float line
> made of weed control cloth
> that looked like a commercial
> fishing net.

FROM THE TOP — Three streamkeepers are
feeding the float line over the side.

>   NARRATOR (V.O.)
>   It had floats on the top and a
>   lead line on the bottom . . .

UNDERNEATH — the dock, the float line is
being pulled between the wrapped pilings.

>   NARRATOR (V.O.)
>   . . . so that the material
>   would stay submerged but still
>   go up and down with the tide.

FROM THE TOP — Matsen and others continue
to pull the float line over the side of the
dock.

>   JONN MATSEN (O.S.)
>   To lay the float line under we
>   need 7, 8 feet of water, and
>   so the tide's about 11, 12 feet
>   now. Within an hour or so it
>   should be low enough we can run
>   it right through.

A STREAMKEEPER — looks at the other
streamkeepers at the south end of the dock.

>   JONN MATSEN (V.O.)
>   Tell them to stop pulling!

>   STREAMKEEPER
>   STOP!!!

ANOTHER STREAMKEEPER — at the far end of the wharf gives the thumbs-up sign.

INT. SQUAMISH TERMINALS OCEAN — NIGHT

A few herring dart around in the dark next to the float line.

CLOSER — on the float line fabric. It's covered with eggs.

> NARRATOR (V.O.)
> Three weeks later, the herring
> responded by laying millions
> of eggs on the new underwater
> surface.

ANOTHER ANGLE — The eggs are in quite a thick layer. They look clean and alive.

MUCH CLOSER — The eggs on the float line are dense, but not too dense.

> NARRATOR (V.O.)
> It appeared the streamkeepers
> had finally provided a safe
> spawning surface for the
> herring.

A FEW HERRING — dart along and past the float line.

ON ANOTHER SECTION — The eggs are much thicker here. It's a concern.

Freezing air temperatures after a spawn at the Squamish Terminals east dock killed most of the eggs laid on the streamkeepers' weed control wraps. This event made the streamkeepers realize that pilings posed a threat to herring eggs not only because of toxins but also because of the risk of air exposure. (Photo: Scott Renyard)

On April 16, 2015, the Squamish Terminals east dock caught fire and the entire structure was destroyed. The streamkeepers' work was also destroyed, and they had to suspend wrapping pilings at the Terminals. (Photo: Eric Andersen)

Soon after the fire, the dock was rebuilt using
different materials. (Photo: Scott Renyard)

The Squamish Terminals management had been monitoring the streamkeepers' findings
and rebuilt the new dock with herring in mind. Herring are known to avoid shiny or
slippery surfaces so that their eggs aren't dislodged by waves or currents. The new pilings
were installed with a shiny black coating over concrete. This strategy appears to have
worked. So far, herring haven't spawned heavily on the new pilings. (Photo: Scott Renyard)

PANNING ALONG — the float line. It ripples like a giant flag in the wind.

> NARRATOR (V.O.)
> And as the float line moved
> back and forth like a giant
> sail, it kept the eggs clean
> and oxygenated in calmer water.

INT. SQUAMISH TERMINALS EAST DOCK OCEAN — ONE WEEK LATER

The eggs on the float line are not doing well. The thick patches are turning brown. Chunks have disappeared. They likely fell to the ocean floor.

> NARRATOR (V.O.)
> But if the spawn was too thick
> or the wave action too great,

TIGHTER — A chunk of eggs has come loose and is about to fall off the material.

> NARRATOR (V.O.)
> . . . many of the eggs were
> dislodged and fell to the
> bottom. It wasn't a perfect
> solution after all.

YET ANOTHER SECTION — has many chunks of eggs about to fall off the float line. And there are obvious patches where eggs are now missing.

FADE OUT:

FADE IN:

OVER BLACK

The chatter of emergency response teams on walkie-talkies.

EXT. SQUAMISH TERMINAL DOCK — DAY

The Squamish Terminals east dock is on fire. Heavy smoke billows into the sky.

SUPERSCRIPT: April 16, 2015.

CLOSER — Flames shoot through the east dock's deck. Black smoke fills the air.

>   NARRATOR (V.O.)
>   A fire broke out at the
>   Squamish Terminals east
>   dock . . .

FROM ACROSS THE WATER — The creosote pilings are on fire. The thick smoke drifts toward Squamish.

>   NARRATOR (V.O.)
>   . . . and it spread quickly
>   through the creosote pilings.

EXT. SQUAMISH STREET — MOMENTS LATER

Black smoke fills the sky.

EXT. MAMQUAM CHANNEL CLIFF — NIGHT

By night, the fire is raging. It's impossible to put out.

>   NARRATOR (V.O.)
>   The dock burned for six days
>   and was completely destroyed.

LATER THAT NIGHT — The scene is apocalyptic. The whole structure is engulfed in flames.

FIREFIGHTERS — try to battle the blaze, but it's futile.

ON THE PILINGS — They are all on fire.

EXT. SQUAMISH NEXEN BEACH — THE NEXT MORNING

The main fire is out. The deck and pilings that remain are black and smouldering.

>   NARRATOR (V.O.)
>   The streamkeepers' work was
>   lost in the blaze, and it
>   seemed all of their efforts
>   were in vain.

TIGHTER — Ironically, the pilings look like fresh creosote pilings, but they are just burnt remnants.

A CRANE — takes large buckets of water out
of the ocean and dumps it on the smouldering
wharf.

FADE TO:

EXT. SQUAMISH TERMINALS EAST DOCK (DRONE) —
DAY

The new dock is finished.

                    NARRATOR (V.O.)
          But sometimes with tragedy,
          something good emerges from the
          ashes.

UNDER THE DOCK — the new structure is clean.
The pilings are concrete, but they're wrapped
in a shiny black coating.

                    NARRATOR (V.O.)
          The managers at the Terminals
          were closely monitoring the
          streamkeepers' project,

TRAVELLING — among the new pilings.

                    NARRATOR (V.O.)
          . . . and used lessons learned
          to help design the new dock.

CLOSER TO THE PILINGS — There is no sign of
pink mould.

                    NARRATOR (V.O.)
          New concrete pilings were
          installed with a shiny, non-
          toxic material. The goal was
          not to protect the eggs from
          the surface of the new pilings,

HERRING — are spawning on algae and not on a
piling.

                    NARRATOR (V.O.)
          . . . but to discourage herring
          from spawning on them in the
          first place.

BACK TO THE PILINGS — A few eggs are
sprinkled on them.

                    NARRATOR (V.O.)
          The next spawning season, only
          a few herring tried to lay
          their eggs on the new pilings.

CLOSER — on a new piling. There are only a
few eggs, and they are dead.

VERY CLOSE — A few sparse patches of goop.

                    NARRATOR (V.O.)
          So far, the strategy has been
          a huge success with only a few
          eggs affected at a place where
          millions used to die each year.

FROM THE SKY — drifting slowly over the new dock.

EXT. FALSE CREEK (DRONE) — DAY

Cruising over False Creek. Fisherman's Wharf is in the foreground.

SUPERSCRIPT: Summer 2013.

>   NARRATOR (V.O.)
>   Prior to the Squamish Terminal
>   fire, the streamkeepers were
>   surveying other marinas in the
>   region . . .

EXT. FISHERMAN'S WHARF — DAY

Jonn Matsen walks down a float. He is looking for signs of herring.

>   NARRATOR (V.O.)
>   . . . to see if pilings were
>   impacting other herring
>   populations.

ON THE SIGN — "Fisherman's Wharf — False Creek * Vancouver"

>   NARRATOR (V.O.)
>   And they discovered a herring
>   hot spot in False Creek,

Prior to the Squamish Terminals fire, the streamkeepers surveyed the region's marinas to see if herring were spawning on other pilings. They found that herring were spawning on pilings at Fisherman's Wharf, which is right next to the area known as Senákw, the former location of the old Squamish First Nation village. (Photo: Scott Renyard)

The floating docks at Fisherman's Wharf move up and down with the tides and are surrounded by boats. These factors created new challenges for the streamkeepers' herring project because wraps made of soft materials would be shredded in days with all the movement. (Photo: Scott Renyard)

Herring often spawn heavily on the Fisherman's Wharf pilings and most of the eggs die, so the streamkeepers decided to wrap the pilings with the same durable plastic used at the Squamish Terminals to see if they could improve herring egg survival on the pilings. (Photo: Scott Renyard)

UNDERWATER — A few herring are spawning on a mussel-covered piling.

> NARRATOR (V.O.)
> . . . right in the heart of
> Vancouver, Canada's largest
> west coast city.

THE HERRING — float gently over the mussels.

> FADE TO:

PHOTO — from 1885. Looking across the entrance of False Creek at the area known as Señákw (or Snawq).

PHOTO — from the Mount Pleasant area looking west. There is just the odd house and two bridges across the water.

PHOTO — of Squamish First Nation men with guns in 1891.

> NARRATOR (V.O.)
> This area was once known as
> Señákw, the winter village for
> the Squamish First Nation.

MAP — of the False Creek and Vancouver area. The old village site lights up in green to highlight its location.

> NARRATOR (V.O.)
> Before Señákw was surrounded by
> a city,

INT. FALSE CREEK OCEAN — DAY

A large sturgeon cruises over a bed of
seaweed.

A LARGE SCHOOL — of salmon swim by a huge
rock.

A SEAL — turns upside down as it plays in
shallow water.

A LARGE SCHOOL — of herring swim by.

          NARRATOR (V.O.)
    . . . it was an area rich with
    large populations of sturgeon,
    salmon, seals and especially
    herring.

ON LARGE HERRING — moving slowly through the
estuary.

          NARRATOR (V.O.)
    But now, those populations are
    a mere fraction of what they
    once were.

PHOTO — of a First Nations family in a dugout
in 1907. They are rowing into the shore near
Kitsilano beach.

MAP — of the Senákw village area. This time
the point is more developed. An animated rail
line is drawn across the old village site.

                    NARRATOR (V.O.)
          In 1886 and 1901, portions of
          the village were expropriated
          to accommodate rail lines.

PHOTO — looking down at Senákw. A small
village has grown up along the shore.

PHOTO — of Senákw. A few bonfires are
smouldering and a few small homes are
scattered about.

                    NARRATOR (V.O.)
          In 1913, the BC Government
          forcibly removed villagers due
          to industrial expansion in the
          area.

PHOTO — looking east from Senákw in 1925.
Smoke stacks and industry line the edges of
False Creek.

PHOTO — of False Creek in 1928. We see a
giant lumber mill and logs covering a portion
of the water.

PHOTO — of a fish processing plant with fish
boats tied up at the dock.

                    NARRATOR (V.O.)
          Over the ensuing decades, this
          bay would be named False Creek
          and became a key waterway for
          Vancouver's heavy industry.

PHOTO — of False Creek in 1942. It is jammed
with lumber mills and floating log storage.

                              DISSOLVE TO:

INT. FALSE CREEK (RE-ENACTMENT) — DAY

The bottom of False Creek is brown, mucky and
filled with log debris.

CABLES — log debris and other waste clog the
sea floor.

              NARRATOR (V.O.)
          Just as in Squamish,
          development destroyed what
          was once a productive estuary.
          And yet, even with more than
          150 years of heavy industrial
          activity,

INT. FISHERMAN'S WHARF OCEAN (2013) — DAY

Herring are spawning on a piling covered in
barnacles.

              NARRATOR (V.O.)
          . . . herring had somehow
          survived and were still
          spawning right next to the old
          village site.

MAP — Pushing in on the south side of False
Creek, the area just east of Senákw.

Arrows point to three creeks that enter the little bay.

> NARRATOR (V.O.)
> This part of False Creek at one
> time had three small creeks.

EXT. FISHERMAN'S WHARF (2019) — DAY

The area seen on the map as it is today. The edges of the water are now banks of riprap. There is a strange current from fresh water bubbling up from a storm drain.

> NARRATOR (V.O.)
> And despite extensive
> development,

CLOSE — on the fresh water bubbling to the surface.

> NARRATOR (V.O.)
> . . . fresh water still bubbles
> out of a storm drain where one
> creek once flowed into the bay.

NEARBY — the herring are spawning on a mussel-covered piling at Fisherman's Wharf.

> NARRATOR (V.O.)
> Some would say, these herring
> are simply drawn to fresh water
> like other herring that spawn
> at the mouths of creeks.

THE HERRING — are spooked and dart away and under the wharves.

ON A WOODEN SCULPTURE — looking out into False Creek at the edge of the water in the area that was once Senákw.

>   NARRATOR (V.O.)
>   But it could be that these
>   herring are specific to this
>   location and . . .

UNDERWATER — A large school of herring swim under the wharves at Fisherman's Wharf.

>   NARRATOR (V.O.)
>   . . . are returning to their
>   birthplace like they have for
>   thousands of years.

BEAUTIFUL HERRING — cruise slowly by, their large eyes scanning the water for signs of danger.

TIGHTER — Herring in a seine net swim erratically in all directions.

INT. OCEAN (70 YEARS AGO) (RE-ENACTMENT) — DAY

A large school of herring swim in a tight school.

ANOTHER LARGE SCHOOL — dart quickly by. They have been spooked by something.

                    DOUG HAY (O.S.)
          One of the things that has
          been done very well in British
          Columbia . . .

RESUME — Hay interview.

                    DOUG HAY
           . . . for, oh, almost 70 years
          now,

A LARGE — school of herring go deeper into
the water.

                    DOUG HAY (O.S.)
           . . . is the tagging of herring.

ANOTHER SCHOOL — in deep water cruise along,
staying close together.

EXT. OCEAN — DAY

An old fishing boat with a lone fisherman
pulling in his net.

                    DOUG HAY (O.S.)
          This started off back in the
          '30s during the so-called
          Reduction Fishery when they
          stuck . . .

POSTER — used by the Government of Canada
to make fishermen aware of the tags and to
remind them to collect them. It has a drawing
of a herring showing the location and design

of the tag. The words "Tagged Herring" are prominent.

> DOUG HAY (O.S.)
> . . . little metal tags in the herring,

A CONVEYOR BELT — pulls herring out of a frothy mixture in a reduction plant.

> DOUG HAY (O.S.)
> . . . and these tags were recovered in the big vats where they reduced the herring down to oil called belly tagging.

IN THE VAT — The herring are gone. Only a frothy mixture remains.

PHOTO — of a small holding pen filled with herring.

PHOTO — of the holding tank. Even though we can't see the herring in the tank, the red spaghetti tags just under the surface are obvious.

> DOUG HAY (O.S.)
> After that, another tagging program was started using Floy tags or little spaghetti tags . . .

Tagged Herring

(location of tags)

The Pacific Biological Station is tagging herring in British Columbia waters to investigate coastal migration routes and mixing of spawning stocks.

The tag is a small red plastic tube implanted on the left side below the dorsal fin as shown above. A $2.00 reward is offered for the return of each tag. Anyone finding a tagged herring is requested to return the tag and information to:

Herring Tags,
Pacific Biological Station,
Nanaimo, B.C. V9R 5K6

For further information contact:
Doug Hay or Carl Haegele
at telephone (collect) 758-5202

Postage paid, self-addressed envelopes for herring tag returns will be available at processing plants and from all Regional Fisheries offices. Recovery information is requested for each tag returned.

For recovery in processing plants, the tag number, plant name and address, date and time of recovery, fish-lot number and in-plant location of recovery are required.

For recovery on vessels, the tag number, vessel name and type, gear type, date and place of capture are required.

Queen's Printer for British Columbia ®
Victoria, 1989

Herring tagging programs began in the 1930s in an effort to better understand the Pacific herring populations on BC's coast. These studies determined that there were likely many distinct herring populations. Some, but not all, herring populations migrate. Some research suggests that inlets often have a resident or local herring population. (Source: Department of Fisheries and Oceans, Canada)

FISHERIES RESEARCH BOARD OF CANADA

ANNUAL REPORT

of the

PACIFIC BIOLOGICAL STATION

for

1941

This 1941 report by R.E. Foerster, a Pacific Biological Station director, summarizes findings from herring vertebrate and tagging studies. Foerster wrote that populations on the east coast of Vancouver Island had "practically no intermingling," whereas populations on the west coast of Vancouver Island were distinct with limited intermingling, and northern populations were likely distinct to at least four geographical areas. More recent studies suggest 25% intermingling of stocks, but this could be because of the current dominance of the east Vancouver Island spawn, which is a migratory population that spends most of its life cycle off the west coast of Vancouver Island. The question is, has this population become dominant because many of the other populations have been reduced or eliminated? (Document: Government of Canada)

PHOTO — of three dead herring on a plank of
wood. The spaghetti tags are visible.

PHOTO — of herring in a small net being
tagged.

PHOTO — of a worker with a tag crimper.

> DOUG HAY (O.S.)
> . . . and more recently there's
> been tagging using nose tags.
> All of this work has been
> written up, and it reveals a
> number of things.

MAP — of the BC coast. First the south
coast lights up, then the north coast to
demonstrate the approximate area covered by
the two zones.

> DOUG HAY (O.S.)
> If you are a herring that's
> tagged in the southern part of
> the coast, you probably stay in
> the southern part. If you're
> tagged in the northern part of
> the coast, you probably stay
> there.

LOOKING STRAIGHT DOWN (DRONE) — Herring swim
very slowly just under the surface.

                    DOUG HAY (O.S.)
          These tagging data also show
          that sometimes herring may not
          move very much.

MAP — of Howe Sound as an example of
residential herring migration.

                    DOUG HAY (O.S.)
          And so, if a fish doesn't move,
          and spends all of its life
          in one area, and if it keeps
          spawning in that area,

INT. HOWE SOUND — DAY

A school of herring cruise over a bed of
bladder wrack kelp.

                    DOUG HAY (O.S.)
          . . . we can incorrectly assume
          that it's going away when in
          fact, it never left in the
          first place.

EXT. HOWE SOUND — DAY

A rockweed shoreline is covered with milt-
laden water. A few herring are emerging and
disappearing from view.

EXT. BC WEST COAST NORTH — DAY

A pristine landscape. The ocean is perfectly still.

                    DOUG HAY (O.S.)
          In many areas of the British
          Columbia coast,

UNDERWATER — Herring dart over rockweed.

                    DOUG HAY (O.S.)
          . . . there's probably herring
          that migrate and herring that
          don't migrate.

DOCUMENT — Taylor, F.H.C. (1964). Life history and present status of British Columbia herring stocks.

ON THE FRONT PAGE — Zoom in on the title then pan down to the date, "1964." Life history and present status of British Columbia herring stocks.

ON PAGE 9 — Pushing in the text. The words "two types" lift off the page, then move to the words "major migratory" and then quickly to "homesteaders."

                    NARRATOR (V.O.)
          Research published in 1964
          determined that there are "two
          types" of herring in British
          Columbia, and they were
          described as "major migratory"
          and minor local populations
          called "homesteaders."

ON A TABLE — Scrolling down a list of herring spawn locations surveyed by the Department of Fisheries and Oceans.

NARRATOR (V.O.)
After recording approximately
"30,000 spawning events" at
"1392 locations,"

MAP — Source: Taylor, F.H.C. (1973). Detailed tagging and tag recovery records of herring, 1957 to 1967, page 31.

NARRATOR (V.O.)
. . . migratory herring on the
British Columbia coast were
determined to be 13 distinct
populations.

DOCUMENT — Foerster, R.E. (1941). Annual report of the Pacific Biological Station for 1941.

ON THE COVER — Moving up to the word "Confidential," then quickly to "1941" at the bottom of the page.

NARRATOR (V.O.)
A once "confidential" report
from 1941 . . .

UNDER A MICROSCOPE — Herring vertebrae of different sizes.

> NARRATOR (V.O.)
> . . . reported that herring
> vertebrae were different in
> number and size from different
> geographical areas.

PAN OVER TO — the words "very definite
segregation." Then the words "significant
differences" leap off the page.

> NARRATOR (V.O.)
> This means that there was
> "very definite segregation"
> and "significant genetic
> differences" between the
> populations.

INT. MILTY OCEAN — DAY

A school of herring are moving like shadows
through the milty water.

MAP — of the western North Pacific. The
ranges of the three types of herring
populations are revealed one by one: marine,
coastal and lake-lagoon.

> NARRATOR (V.O.)
> Across the ocean, scientists
> classified herring into three
> distinct types in the west
> North Pacific: marine, coastal
> and lake-lagoon.

THE RANGE — of marine herring is highlighted.

>   NARRATOR (V.O.)
>   Marine herring migrate large
>   distances and spend their
>   entire lives in high-saline
>   ocean waters.

THEN THE RANGE — of coastal herring is
highlighted with a different colour.

>   NARRATOR (V.O.)
>   Coastal herring inhabit inlets
>   and bays and usually do not
>   migrate long distances.

LASTLY THE RANGE — of lake-lagoon is
highlighted with yet another colour.

>   NARRATOR (V.O.)
>   And lake-lagoon herring spend
>   most of their lives in brackish
>   water and migrate to adjacent
>   marine water for food.

WATCHING — herring spawning in the shallows.
They flip out of the water in a frenzy.

>   NARRATOR (V.O.)
>   But defining individual herring
>   populations is difficult at
>   best.

SMALL WAVES — push a few spawners up onto
the shore. They scramble to get back into the
water.

ONE SPAWNER — wriggles back into the water.

>   NARRATOR (V.O.)
>   New research on Canadian
>   Pacific herring reveals that
>   intermingling or straying
>   between herring stocks reaches
>   up to 25 percent.

SEVERAL HERRING — work hard to stay in the water when a small wave pushes them up onto the shore.

WIDER — Herring leap up and flip out of the water near the shore.

>   NARRATOR (V.O.)
>   This straying reduces the
>   genetic distinctiveness between
>   herring populations, and this
>   causes a tremendous amount of
>   uncertainty when it comes to
>   fisheries management.

EXT. SALISH SEA — DAY

Commercial fishing vessels are spread over a milty part of the ocean.

CLOSER — A gillnetter is harvesting a few herring.

                    DAVID ELLIS (O.S.)
          Al Hursten adamantly wanted
          smaller quotas and no fishing
          until . . .

RESUME — Ellis interview.

**LOWER THIRD: David Ellis, Former Head,
Pacific Fisheries, COSEWIC (Committee on the
Status of Endangered Wildlife in Canada).**

                    DAVID ELLIS (V.O.)
          . . . there's adequate spawn
          for each local stock, and he
          defined herring as locally
          recruiting . . .

IN LAMBERT CHANNEL — herring fishermen are
working their nets.

                    DAVID ELLIS (V.O.)
          . . . or locally homing stocks.

FROM SHORE — Several boats are spread out
along the edge of the milty water.

                    DAVID ELLIS (V.O.)
          He knew what was going on. He
          wanted small quotas for each
          inlet.

INT. GENETIC TESTING LABORATORY — DAY

Herring are laid out on a tray.

A SCIENTIST — preps a solution and takes a sample from one of the fish.

                    NARRATOR (V.O.)
          Herring bones, collected during
          the 2013 study,

UNDER — a dissecting microscope. The scientist teases apart the bones of a herring spine.

                    NARRATOR (V.O.)
          . . . were analyzed using a new
          microsatellite genetic test.

UNDERWATER — A herring school swims by. The herring are somewhat dispersed, rather than swimming close together.

EXT. SALISH SEA — DAY

The coast stretches out as far as the eye can see.

                    NARRATOR (V.O.)
          This test found that 43
          distinct herring populations
          once existed along the British
          Columbia coast.

A HERRING BALL — spins and twists as the herring try to fend off an attack from marbled murrelets.

> NARRATOR (V.O.)
> This likely means that, at
> one time, nearly every inlet
> on the coast had a resident
> population.

LOOKING UP — The herring spin in a ball like
a tornado.

A DIFFERENT SCHOOL — of herring swim off into
a green abyss.

> NARRATOR (V.O.)
> This also means that herring
> were far more abundant and
> geographically variable prior
> to industrial fishing pressure.

MORE HERRING — swim in a dirty green ocean.

ANOTHER SCHOOL — swim quickly in blue-green
water.

> NARRATOR (V.O.)
> Herring can take hundreds,
> if not thousands, of years
> to evolve into distinct
> populations.

EXT. OCEAN (DRONE) — DAY

A cluster of commercial fishing boats are
tied up together.

LOOKING DOWN — Hundreds of birds are flying around two seiners and a skiff.

>   NARRATOR (V.O.)
>   The fear is that our fishing
>   practices have been hastily
>   and poorly executed, and that
>   genetic diversity has been
>   lost.

INT. OCEAN (MANY DECADES AGO) — DAY

A massive school of herring from a long time ago, much larger than anything that has been seen in modern times, weave through a bull kelp forest.

>   NARRATOR (V.O.)
>   Ignoring that the British
>   Columbia coast was home
>   to many distinct herring
>   populations . . .

EXT. WEST COAST OCEAN (RE-ENACTMENT) — DAY

A large seiner begins to retrieve its massive net.

>   NARRATOR (V.O.)
>   . . . allowed the reduction
>   and sac-roe fisheries to
>   aggressively catch herring,

THE NET — is nearly fully drawn in. It is loaded with tons of herring.

                    NARRATOR (V.O.)
          . . . which may have caused
          the extinction of some smaller
          stocks.

TIGHTER — Thousands of herring form a
wriggling mass in the net.

FROM THE SIDE — The herring are a solid mass
of fish.

                    NARRATOR (V.O.)
          The trouble is, what stocks
          have been eliminated and what
          effect has this had on the
          marine ecosystem?

UNDERWATER (1970s) — The herring school seems
endless.

                    NARRATOR (V.O.)
          The loss of herring populations
          with different spawn timings,

INT. BIG QUALICUM RIVER (UNDERWATER) — DAY

Chinook and coho salmon migrate up the river.

                    NARRATOR (V.O.)
          . . . is likely to have
          profound effects on salmon
          populations.

BUBBLES — fill the view as a large chinook
salmon appears and then disappears upstream.

INT. BC INLET — LATE SPRING

A small school of salmon smolts are heading
out to the ocean.

SMOLTS — make their way out into the ocean.
They are thin and wide-eyed.

> NARRATOR (V.O.)
> Each spring and summer, young
> salmon migrate to the ocean
> looking for food.

ELSEWHERE — only a few salmon smolts swim
through frame.

> NARRATOR (V.O.)
> But if young herring are
> absent, salmon smolts have to
> look harder for prey and fewer
> will survive.

MAP — of the BC coast showing the general
plankton bloom as it moves from south to
north during the spring.

> NARRATOR (V.O.)
> And there is even more
> complexity when it comes to
> herring spawning on the Pacific
> coast. Each wave of spawning
> doesn't happen coast-wide at
> the same time. It's staggered
> and generally moves south to
> north as the sun triggers
> plankton blooms each spring.

ON THE SHORE — a bald eagle grabs a herring
in its talons. Then it picks it up in its
mouth and gulps it down.

> DOUG HAY (O.S.)
> It could be the absence of a
> number of predators such as a
> concentration of birds,

NEARBY — a pair of sea lions surface. One
grabs a herring and swallows it.

A SEA OTTER — has a herring by the tail.

> DOUG HAY (O.S.)
> . . . or a concentration of
> mammals.

RESUME — Hay interview.

> DOUG HAY
> On the other hand, it has to be
> the right combination of water
> temperature, salinity up to a
> point, because herring . . .

ALONG THE SHORELINE — herring flip in the
shallow water.

> DOUG HAY (O.S.)
> . . . by and large don't spawn
> in fresh water.

MILTY WATER — rolls over a bed of rockweed.
Herring are wriggling in the algae.

>   DOUG HAY (O.S.)
>   The elements in the marine
>   environment that would lead a
>   herring to spawn . . .

LOOKING — along the shoreline. There are no
predators, no birds. Just herring.

>   DOUG HAY (O.S.)
>   . . . may be the absence of
>   things rather than the presence
>   of things.

A ROCKY OUTCROP — Sea plants are covered in
fresh herring eggs.

UNDERWATER — The marine plants are thick with
eggs.

CLOSER — Eggs on rockweed are clean and alive
and have visible yolks.

>   FADE OUT:

FADE IN:

EXT. FISHERMAN'S WHARF — NEW YEAR'S EVE

Establish.

JONN MATSEN — is giving instructions to the
other streamkeepers and the volunteers.

                    JONN MATSEN
          I'm Jonn Matsen. I'm the
          coordinator.

ON DOUG SWANSTON — He nods.

                    JONN MATSEN (O.S.)
          Douglas Swanston is the, the
          contractor on this project.

BACK TO MATSEN

                    JONN MATSEN
          If they spawn too high out of
          the water, they're going to die
          from the sun, frost or wind is
          going to get them, right?

ON THE GROUP — They are listening intently.

                    JONN MATSEN (O.S.)
          So we are actually thinking,

BACK TO MATSEN

                    JONN MATSEN
          . . . maybe we will make some
          skirts to put over the eggs
          after they hatch so it dangles
          down sort of like a kelp would
          to protect them.

ON SWANSTON

LOWER THIRD: Doug Swanston, Biologist, NW Seacology.

SWANSTON — shows the others the wrap material and the different textures on its front and back.

> DOUG SWANSTON
> The material is two-sided,
> there is a smooth side. You can
> see it, can everyone see that
> in the light? And then there's
> a rough side.

TIGHTER — as Swanston describes the plan.

> DOUG SWANSTON
> About a half an inch from the
> edge at the top and the bottom,
> near the top and bottom, we put
> a nail in just to hold it in
> place, while we wait for the
> tide to drop and come up here
> to have our little ciders . . .

OVER THE SHOULDER — of Patrick MacNamara. He is listening carefully.

> DOUG SWANSTON (O.S.)
> . . . and water and whatever
> to celebrate. Right?

BACK TO MATSEN

                    JONN MATSEN
          It's going to be fun. It's
          always an adventure. We never
          know what we're going to see.

EXT. FISHERMAN'S WHARF (2013) — NIGHT

The crew make their way down the ramp.

ON MACNAMARA — scraping barnacles off a
piling to clear the way for the wrap.

                    NARRATOR (V.O.)
          As the clock approached
          midnight on New Year's Eve, the
          tide was finally low enough
          for the streamkeepers to begin
          wrapping the pilings.

ON ROY SAKADA — using a shovel to clean off
the barnacles.

                    NARRATOR (V.O.)
          The hard plastic wraps that
          worked well in Squamish would
          prove . . .

TWO MEN — put the wrap into place on a
piling.

                    NARRATOR (V.O.)
          . . . even more important with
          the floating wharves at this
          location.

A ROPE — is placed around the wrap to secure it around the piling.

>   NARRATOR (V.O.)
>   These wharves bang constantly
>   against the pilings with the
>   ebb and flow of the tides and
>   would certainly shred a soft
>   material in mere days.

A PAIR OF HANDS — nail a wrap to a piling with a hammer.

INT. FISHERMAN'S WHARF OCEAN — SIX WEEKS LATER

A wrapped piling is swarmed by spawning herring as they lay their eggs. It's an amazing sight.

SUPERSCRIPT: February 14, 2014.

>   NARRATOR (V.O.)
>   Six weeks later, on Valentine's
>   Day, herring spawned heavily on
>   the new wraps.

LOOKING DOWN — the piling at the frantic spawning activity.

FROM THE SIDE — The herring rub their bellies on the piling in all directions.

This is an example of a Fisherman's Wharf piling with a fresh wrap installed by the Squamish Streamkeepers. It has fresh eggs on it, and the eggs in this photo are still alive. (Photo: Scott Renyard)

Eggs on the newly wrapped pilings initially looked healthy, and the streamkeepers were shocked when most of them died. At first they thought the deaths were a result of exposure to air. They later discovered that oil and fuel floating on the water's surface coat the herring eggs, which are stationary on the pilings, as the tides go up and down each day. (Photo: Scott Renyard)

The streamkeepers modified the float line concept used in Squamish. This time they used fine netting so that water could pass through it. This modification was done with the hope that the float line wouldn't move around as much in tidal currents. There was a real concern that it could get tangled in boats moored at the docks, which could put a stop to the enhancement work. They also decided to place it along the shoreline as an extra precaution to avoid upsetting boaters. This is a close-up of the eggs herring laid on the float line. (Photo: Scott Renyard)

The mesh float line used at Fisherman's Wharf became clogged with brown algae because it was exposed to the sun. Herring spawned on it when it was clean, but avoided it during subsequent spawning waves. (Photo: Scott Renyard)

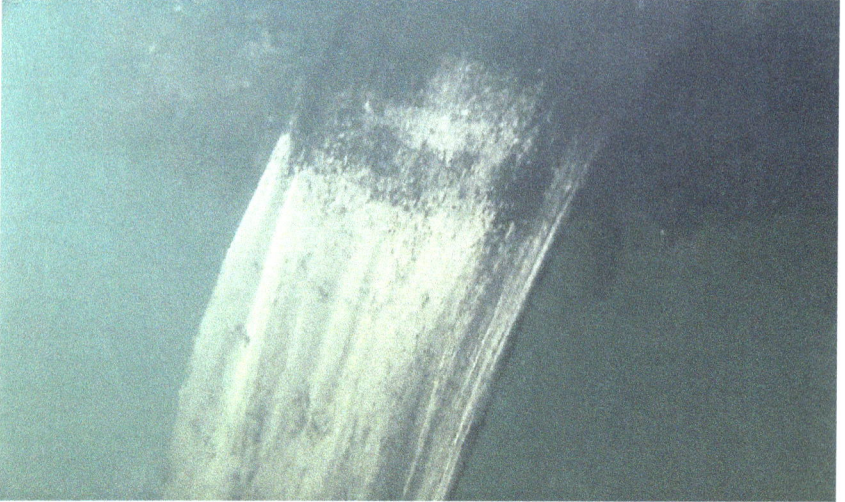

The new mesh panels allow water to flow through them, but they also catch silt and other debris. The silt in the water at False Creek clogged the nets, and the streamkeepers needed to shake the panels every few days while they waited for herring to come and spawn on them. (Photo: Scott Renyard)

Herring spawn heavily on the panels placed in False Creek. These panels keep the eggs submerged because they are attached to the floating docks and go up and down with the tide. This protects the eggs from the creosote on the pilings and from being exposed to the air, which could dry or freeze them. It also keeps them underwater and away from oil contaminants floating on the surface of the water. (Photo: Scott Renyard)

FRANTICALLY — they lay their eggs, building in intensity as they try to lay them before being chased or scared away by predators.

THE HERRING — are now all around the lens of the camera.

INT. FISHERMAN'S WHARF OCEAN — THE NEXT DAY

A piling is covered in clean, white herring eggs.

ON A SECTION — of the piling. The eggs still look spectacular.

> NARRATOR (V.O.)
> And for a few days, the eggs
> looked beautiful and healthy.

EXT. FISHERMAN'S WHARF — A FEW DAYS LATER

The eggs on the pilings are now dead and turning brown.

TIGHTER — Patches of the dead eggs are turning to slime.

EVEN TIGHTER — The eggs are covered in a white mould.

> NARRATOR (V.O.)
> But as the days passed, many of
> the eggs began to die.

SOME EGGS — now look more like stucco than eggs.

                    NARRATOR (V.O.)
          This time, however, it wasn't
          from exposure to creosote or
          extreme air temperatures.

EXT. FALSE CREEK SHORELINE — DAY

A blue sheen of oil streams from the shallows.

                    NARRATOR (V.O.)
             It was from exposure to oil.

INT. FALSE CREEK OCEAN — EARLIER

The herring are spawning on a piling. Radio chatter.

A SAILBOAT — is on fire in False Creek.

A THICK OIL SHEEN — swirls near Fisherman's Wharf.

A BOAT — tied up to a wharf is engulfed in flames.

HERRING — are spawning on a piling.

BACK TO THE LAST BOAT — It continues to burn.

THE OILY SHEEN — is thicker between some boats.

ANOTHER BOAT — smoulders away.

A PILING — is covered in fresh eggs.

MORE SMOKE — fills the marina as another boat burns.

THE OIL SLICK — swirls.

ANOTHER BOAT — is engulfed in smoke.

MORE OIL — is on the water.

HERRING — continue to spawn.

ANOTHER BOAT — is on fire, contaminating False Creek water.

MORE OIL — floats on the water.

PANNING ACROSS A PILING — The eggs are not clean.

ANOTHER BOAT ON FIRE — Smoke covers the bay.

OIL — surrounds a piling and touches the eggs.

MORE OIL — floats on the surface of the water.

OLD EGGS ON A PILING — Some have died. Those that have survived look to be in rough shape.

OTHER EGGS — Some are dead. Those that are still alive have oil sticking to their surfaces.

HERRING — oblivious to the dangers, continue to lay their eggs on the pilings.

                                    FADE OUT:

FADE IN:

INT. FALSE CREEK OCEAN — DAY

A wrapped piling is covered in fresh eggs.

                    NARRATOR (V.O.)
            When the streamkeepers realized
            that the eggs on False Creek
            pilings were in constant
            jeopardy,

UNDERWATER — A float line is stretched out under the water.

                    NARRATOR (V.O.)
            . . . they decided to repeat
            their Squamish float line
            strategy and keep the eggs
            submerged.

A BOAT PROPELLER — starts up.

> NARRATOR (V.O.)
> But there was a substantial
> risk that a float line in this
> location would get tangled in
> boats.

LOOKING DOWN (DRONE) — One wharf runs
parallel to the shore and is away from the
boats moored at other floats.

FROM THE GROUND — One spot next to the shore
is free of boats.

> NARRATOR (V.O.)
> So they decided to locate the
> float line along the shoreline
> and away from the boats.

A FINE MESH NET — is dropped into the water.

> NARRATOR (V.O.)
> And also to make the curtain
> out of fine netting so that
> water could pass through it to
> reduce its movement with the
> shifting tides.

THE CURTAIN — is hanging underwater. Herring
are spawning on the net.

CLOSER — The herring like the netting and are
spawning on it.

                    NARRATOR (V.O.)
          And it took just two weeks
          before a large school of
          spawners found the float line
          and covered it with eggs.

LOOKING ALONG — the net as the herring lay
eggs on the mesh.

A LITTLE LATER — The herring are gone and the
mesh is covered in fresh eggs.

VERY CLOSE — The eggs are so thick they fill
the spaces between the mesh.

TWO WEEKS LATER — The eggs on the mesh are
alive. We can see the eyes of the growing
embryos.

SUPERSCRIPT: Two weeks later.

ANOTHER MESH PANEL — The eggs are even
denser, and they all have embryos. Some of
the embryos twist inside the eggs.

INT. FISHERMAN'S WHARF SHORELINE — DAY

The float line hangs perfectly from the top
rope, but it looks nothing like the curtain
that was deployed just a couple of weeks ago.
It's covered in a brown slimy algae.

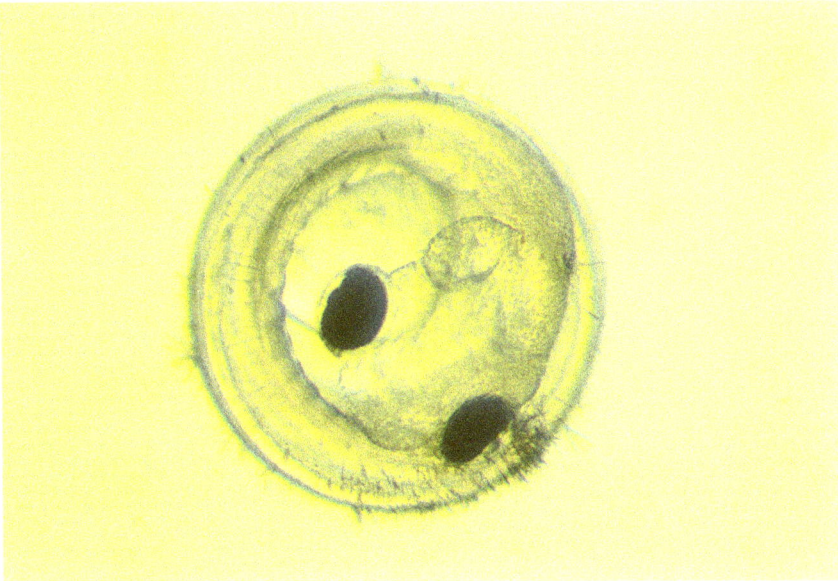

The herring egg's wall is so thin that the embryo can be seen through the shell. Malformations occur when herring eggs and their embryos are exposed to oil or oil products. It's easy to see why herring eggs are vulnerable to oil when the eggshells are so soft and transparent. (Photo: Doug Hay)

LOOKING STRAIGHT DOWN — the float line as it
disappears in the silty green water. If we
didn't know better, we might think the float
line had been hanging here for months.

>   NARRATOR (V.O.)
>   The eggs seemed to survive. But
>   unlike the Squamish Terminals
>   location, this curtain was
>   exposed to sunlight,

APPROACHING THE TOP ROPE — The algae has
covered the rope and the mesh, and barely any
water can pass through the mesh.

>   NARRATOR (V.O.)
>   . . . and it was soon covered
>   in algae so thick that the next
>   wave of spawners avoided using
>   it.

MOVING ALONG — the float line. The algae has
plugged the mesh.

A SCHOOL OF HERRING — swimming under the
wharf turn toward the pilings.

ON A PILING — Herring are spawning on an
unwrapped piling.

                    NARRATOR (V.O.)
          And tragically, the herring
          went back to the pilings to
          spawn.

INT. UNDER THE WHARF OCEAN — DAY

A boat is tied up to the wharf.

BACK TO THE HERRING — They are spawning on
the unwrapped piling.

LOOKING UP — from under the wharf. Boats
cover the surface next to the walkway.

                    NARRATOR (V.O.)
          The next fix was to make
          smaller panels that could be
          tucked around the boats . . .

BACK TO — the herring spawning on a barnacle-
covered piling.

A BEAUTIFUL SHOT — of a few herring laying
eggs on a piling.

                    NARRATOR (V.O.)
          . . . with the hope that the
          herring would ignore the
          pilings and lay their eggs on
          the mesh.

EXT. PARKING LOT — DAY

The streamkeepers are manufacturing more spawning panels. The panels are made of fine mesh netting approximately 2.5 metres long by 1.2 metres wide, with a lead line on the bottom and a piece of white 1-inch PVC piping with a string attached to hang it from the wharf. Zap straps are used to secure the net to the pipe and lead lines.

A VOLUNTEER — is attaching the pipe to the top of the panel.

HANDS ARE BUSY — making the new panels and fastening the mesh with zap straps.

ON JONN MATSEN — making a panel.

ON A WHARF — A panel is deployed by tossing the lead line into the water.

UNDERWATER — The panel drifts down into the water.

WIDER — Three panels hang in the water next to a piling.

                    NARRATOR (V.O.)
          But as the days and weeks
          passed,

MATSEN — tosses a panel into the water.

                    NARRATOR (V.O.)
          . . . the streamkeepers began
          to wonder if they'd missed the
          last wave of spawners for the
          season.

UNDERWATER — A net drifts down through the
water column.

FROM BELOW THE WHARF — The spawning panels
are interspersed between the pilings.

SOME PANELS — have zap straps fastened around
them to shorten them. The streamkeepers
discovered they were too long and dragged in
the mud at the bottom of the ocean.

                    NARRATOR (V.O.)
          And as the panels hung there
          like strange aquatic sculptures,
          another problem emerged.

A PANEL — is lying almost flat in the water
at the top as though it were made of solid
material. It's actually mesh filled with
silt.

                    NARRATOR (V.O.)
          Silt clogged the nets so bad
          that the tidal currents pushed
          them around as if they were
          made of solid material.

A ROCK RAKE — enters the frame and pulls on the silt-laden panel, shaking it off the piling.

THE SILT CLOUD — fills the view like an underwater dust storm.

THE PANEL — is released and plunges straight back down into the water.

WIDER — The silt cloud spreads out and obscures most of the dock and panels.

INT. FISHERMAN'S WHARF OCEAN — A FEW DAYS LATER

Herring are spawning on the panels. A large school approaches a row of panels.

HERRING — are swarming the panels.

WIDER — The herring continue to spawn on the panels.

                    NARRATOR (V.O.)
          Just about when everyone had
          given up, the herring arrived
          and spawned heavily on the new
          panels.

EXT. FISHERMAN'S WHARF — DAY

Patrick MacNamara and Jack Cooley adjust a panel that has just been cleaned of silt.

UNDERWATER — A panel is covered in eggs.

CLOSER — The eggs look fresh and alive.

MORE EGGS — These eggs are alive and have viable yolks.

> NARRATOR (V.O.)
> All of the hard work paid off and millions of eggs were saved.

EXT. FISHERMAN'S WHARF — DAY

A seagull pecks at the eggs on a piling.

INT. FISHERMAN'S WHARF — DAY

A seal cruises under the wharf. A spawning panel can be seen hanging in the background.

> NARRATOR (V.O.)
> And incredibly, this compromised bay came to life with a range of species that rely on herring for food.

A WHARF — hovers over top with a piling and a mesh panel nearby. A school of sticklebacks cruise around the net.

A CRAB — clings to the bottom of a panel. It's eating eggs.

ANOTHER CRAB — is pulling off chunks of eggs from another panel.

A THIRD CRAB — is using its claws to grab the eggs.

A DUCK — swims underwater toward a panel and eats eggs off it.

STICKLEBACKS — hover around a piling and peck away at the eggs.

A SEAL — cruises down a piling with a thick coating of herring eggs and scares the spawners away.

ANOTHER ANGLE — The seal slides down a panel, head first. Its face is sprinkled with herring eggs.

MOMENTS LATER — the seal, still upside down, grabs the panel with its front flippers. The herring are busy spawning on the bottom of the panel, seemingly unaware of the danger.

ONE HERRING — gets too close. The seal grabs the herring with its mouth and eats it.

ANOTHER SEAL — uses the same technique to catch and eat a herring.

EXT. FISHERMAN'S WHARF BAY — MOMENTS LATER

The two seals surface to take a breath.

NEAR THE SHORE — river otters leap into the
water.

>   NARRATOR (V.O.)
>   It's clear that the
>   streamkeeper project is making
>   a difference in False Creek,
>   but is it significant in a
>   larger context?

FROM THE SKY (DRONE) — looking down over the
Fisherman's Wharf area.

                                        FADE OUT:

FADE IN:

INT. DEEP OCEAN — DAY

A large school of herring are bathing in the
deep green ocean. They are scattered, not
staying close together.

FROM LOWER DOWN — The school moves over a
rocky ocean bottom. There are thousands of
them.

THE HERRING — seem to be sprinkled throughout
the deep water.

>   DOUG HAY (O.S.)
>   When we looked at the herring
>   data from British Columbia
>   we've seen that they,

RESUME — Hay interview.

>   DOUG HAY
>   . . . at any one time, have
>   used about 5,000 kilometres.

BC SPAWN RANGE MAP — shows what 5,000
kilometres of spawn would look like on the BC
coast. False Creek and Squamish light up to
show how small these areas are in the context
of the total area used by herring on the
coast.

>   DOUG HAY (O.S.)
>   So about one-fifth of the whole
>   coastline at any one time or
>   other . . .

FROM ABOVE (DRONE) — Herring milt encircles a
small island.

>   DOUG HAY (O.S.)
>   . . . has been used for
>   spawning, which is remarkable.

THEN UNDERWATER — A few herring dart over
rockweed covered in eggs.

>   DOUG HAY (O.S.)
>   In a year when herring are
>   really quite abundant,

RESUME SPAWN RANGE MAP — The 5,000 kilometres
morph to just 500 kilometres of habitat used
in a typical year.

> DOUG HAY (O.S.)
> . . . the cumulative length of
> herring would be around 500
> kilometres.

UNDERWATER — The herring are cruising along
the shore in milty water.

> DOUG HAY (O.S.)
> So what that means is that they
> use only about 10 percent of
> the available spawning habitat.

EXT. LARGE WEST COAST SHORELINE (DRONE) — DAY

The shoreline is covered in milt.

> DOUG HAY (O.S.)
> In other words,

LOOKING STRAIGHT DOWN — A herring skiff
motors over a patch of clear water near milty
waters.

> DOUG HAY (O.S.)
> . . . spawning habitat is not
> necessarily limiting in British
> Columbia.

EXT. ANOTHER SHORELINE (DRONE) — DAY

The water is thick with milt.

LOOKING DOWN — A sea lion bobs about in
thick, milty water.

                    DOUG HAY (O.S.)
          There's probably several
          hundred kilometres that we
          have . . .

ELSEWHERE — herring are swimming in a large
school along a shoreline.

                    DOUG HAY (O.S.)
          . . . identified as being core
          herring spawning habitats.

EXT. MAMQUAM BLIND CHANNEL ENTRANCE — DAY

A giant log sort extends into the water with
several log booms.

                    DOUG HAY (O.S.)
          And as much as possible we
          should resist development of
          the coastline . . .

RESUME — Hay interview.

                    DOUG HAY
          . . . for industrial purposes
          or perhaps for other
          aquacultural purposes in those
          areas.

EXT. HOWE SOUND LOG DUMP — DAY

A load of logs are dumped into the ocean.

EXT. SQUAMISH TERMINALS EAST DOCK — DAY

Jonn Matsen walks down the wharf. A float
line is stretched out on the deck.

> NARRATOR (V.O.)
> The question is, are the
> streamkeeper projects in
> Squamish and False Creek
> located in core habitat?

INT. FISHERMAN'S WHARF OCEAN — DAY

Herring are spawning on two panels hanging in
the water.

THREE SHOTS — of herring spawning on the
panels.

> NARRATOR (V.O.)
> Year after year, spawning
> occurs in these locations
> in spite of the tremendous
> alterations to the environment.

FROM THE SIDE — Herring are spawning on both
sides of a panel at the same time.

> NARRATOR (V.O.)
> This fact alone suggests that
> these places are essential for
> herring.

UNDERWATER — Herring swim quickly along the
bottom of the ocean.

ONE HERRING — leaves the panel where it has been spawning to join the fray at another panel.

EXT. LAMBERT CHANNEL — DAY

A sea lion swims through milty water.

>   NARRATOR (V.O.)
>   At a 2014 Salish Sea ecosystem
>   conference, presenters also
>   reported that herring spawns
>   have completely disappeared
>   from . . .

MAP ANIMATION — moving quickly from location to location, which makes it difficult to determine how far apart these locations are until we pull back to reveal . . .

THE DISTANCE — and as the labels pop up, we realize that the locations are spread over the whole BC coast. This problem is clearly coast-wide.

>   NARRATOR (V.O.)
>   . . . Clayoquot Sound, Barkley
>   Sound, Cowichan Bay, Genoa Bay,
>   Porpoise Bay, Gorge Harbour,
>   Spiller Channel, Skidegate and
>   many more.

HERRING — spawning and jumping along a shoreline.

At a Salish Sea ecosystem conference, First Nations presenters stated that herring spawning activity had disappeared from places that had annual spawns as far back as anyone could remember. This map illustrates that these disappearances have occurred over the entire range of the BC coast. (Map: Juggernaut Pictures Inc.)

                    NARRATOR (V.O.)
          Should all of these areas be
          considered core habitat for
          herring even though they are
          gone?

A GILLNETTER — harvests a few herring.

                    NARRATOR (V.O.)
          And are there any clues as to
          what really caused this to
          happen?

HERRING — are dragged onto the back of the
gillnetter in a net.

                    NARRATOR (V.O.)
          Was it commercial fishing gone
          wrong?

ALONG A SHORELINE — Old creosote pilings are
remnants of an old wharf.

                    NARRATOR (V.O.)
          Was it spawning habitat
          degradation?

BUBBLES AND RAYS OF LIGHT — create a magical
scene as diving birds and dolphins chase
herring through the water.

                    NARRATOR (V.O.)
          Or was it from some other
          factor?

A BALL — of herring try to avoid predator
birds.

>   NARRATOR (V.O.)
>   What we do know is that herring
>   populations off the coast of
>   British Columbia have been
>   acting abnormally for the last
>   three decades.

EXT. SALISH SEA (DRONE) — DAY

A spectacular view of herring milt along a
Hornby Island shoreline.

ANOTHER SHORELINE — Sporadic patches of milty
water create a fascinating mosaic of colours.

>   NARRATOR (V.O.)
>   The Department of Fisheries
>   and Oceans has claimed that BC
>   herring is genetically uniform
>   coast-wide.

EXT. HORNBY ISLAND (DRONE) — DAY

A major spawn is occurring around the island.
A large school of herring swim from clear
water to milty water.

>   NARRATOR (V.O.)
>   And the only significant change
>   has been a concentration of
>   spawning along the east coast
>   of Vancouver Island,

TIGHT TO A SHORELINE — Milt-laden water is calm until herring suddenly burst out of the water near the shore.

> NARRATOR (V.O.)
> . . . since the overall biomass has remained roughly the same.

UNDERWATER — The murky green water makes the school of herring look like shadowy ghosts.

> NARRATOR (V.O.)
> But this is not the way herring have traditionally spawned on the British Columbia coast.

UNDERWATER — Herring are swimming in the milt.

> NARRATOR (V.O.)
> The pattern actually suggests that many herring populations have been reduced or eliminated.

A SCHOOL — of herring swim through a ray of sunlight.

> NARRATOR (V.O.)
> This has allowed the size of the Vancouver Island east coast population to go up.

ANOTHER SCHOOL — of herring turn and swim in a different direction.

HERRING — struggle in thick rockweed at the shore's edge.

>NARRATOR (V.O.)
>At first blush, this might not seem to be a big deal. But it's likely a really big deal.

FROM THE OCEAN FLOOR — A herring ball hovers overhead.

>NARRATOR (V.O.)
>It means that whatever caused the collapse of the other herring populations is lurking nearby, and this population might be next.

HERRING — swim quickly in a dark ocean.

>NARRATOR (V.O.)
>And in fact, this appears to be happening right now.

EXT. LAMBERT CHANNEL — DUSK

A lone gillnetter is harvesting some herring.

CLOSER — Herring are being pulled up onto the deck.

The Reduction Fishery Era lasted from 1937 to 1967, when herring catches went over 250,000 tons. The herring stocks crashed as a result, forcing the Canadian Government to shut down the fishery. When it reopened in the early 1970s, the new policy was to restrict the catch to 20% of the estimated biomass. (Source: City of Vancouver Archives)

In 2022, the DFO reduced the commercial catch to just 10% of the estimated biomass because herring stocks continued to decline. (Photo: Juggernaut Pictures Inc.)

Cherry Point herring population

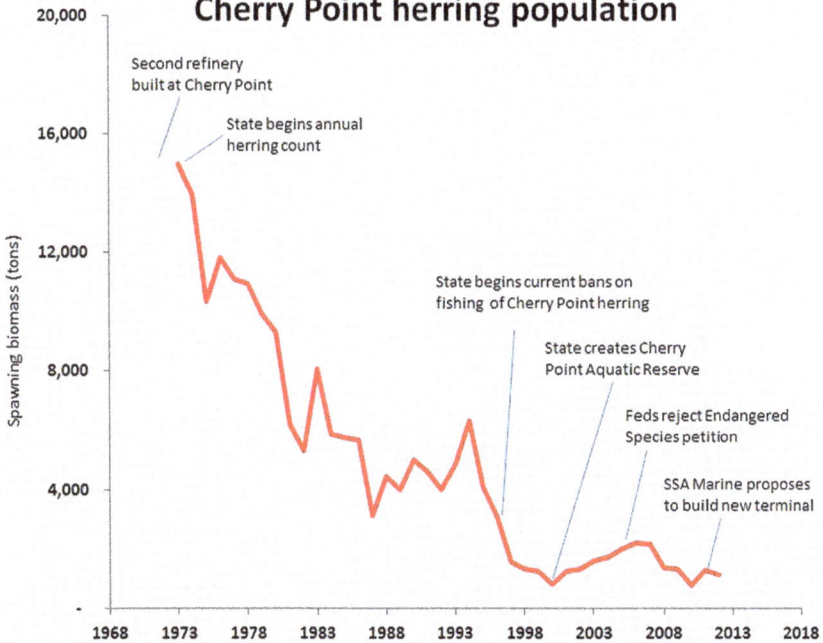

Source for herring data: Kurt Stick, Washington Department of Fish and Wildlife

Research at the Washington Department of Fish and Wildlife found that the herring population in Puget Sound known as the Cherry Point herring had declined by 90%. But the Squamish Streamkeepers have recently observed late herring spawns in False Creek and Howe Sound that match the late spring spawn timing of the Cherry Point herring population. So, even though this population has been reduced at Cherry Point, there is hope that the April spawners are part of the Cherry Point population and have a wider range than previously thought. This means that this genetic strain is still thriving, and there is hope that it can rebound to previous levels of abundance. (Graph: Kurt Stick, Washington Department of Fish and Wildlife)

> NARRATOR (V.O.)
> In 2022, the Canadian
> Government had to reduce the
> commercial catch to just 10
> percent of the estimated
> biomass . . .

A FEW HERRING — swim swiftly by.

> NARRATOR (V.O.)
> . . . because so few fish
> appeared at the spawning
> grounds.

A SEAL — swims in pursuit of a few herring.

EXT. CHERRY POINT — DAY

Kurt Stick, Washington Department of Fish
and Wildlife, is steering a Zodiac along the
shore.

> NARRATOR (V.O.)
> Mysterious drops in herring
> biomass didn't just start
> happening in the last few
> years.

IN ANOTHER BOAT — Fred Felleman and crew are
preparing to take samples to see if herring
have spawned in the area.

STICK AND FELLEMAN — are discussing the
operation.

> NARRATOR (V.O.)
> This phenomenon began in the
> late 1970s.

GULLS — are landing nearby, which suggests
the presence of a few herring.

> NARRATOR (V.O.)
> One of the first populations
> to collapse, the Cherry Point
> herring in Puget Sound,

ONE GULL — has found a herring and gobbles it
down.

> NARRATOR (V.O.)
> . . . has suffered a nearly 90
> percent decline since 1977.

FELLEMAN — points in the direction of the oil
refinery and shipping dock.

RISING ABOVE (DRONE) — the trees to reveal a
large oil refinery at Cherry Point.

> FRED FELLEMAN (O.S.)
> We have two refineries and an
> aluminum smelter all in this
> area, all having discharge
> pipes. Metric tonnes of oil are
> allowed to be released a year
> legally.

INT. HOTEL ROOM (GREEN SCREEN) — DAY

Establish. Interview.

Green screen plate: A portion of a Cherry Point beach. Fred Felleman, a herring activist, later became a key figure at the Port of Seattle.

LOWER THIRD: Fred Felleman, Environmental Consultant, Seattle, Washington.

>   FRED FELLEMAN
>   And so needless to say, this
>   stock of herring now has become
>   unique not just in its run
>   timing . . .

TRAVELLING — along an oil tanker docked at the oil refinery shipping terminal, then swinging over to see another oil tanker.

>   FRED FELLEMAN (O.S.)
>   . . . but also in its
>   proportion of genetic
>   malformities.

VERY CLOSE UP — Herring larvae spin inside eggs covered in oil.

HERRING EGGS (MACRO) — are covered in oil sludge.

                    NARRATOR (V.O.)
          Malformities are common when
          eggs are exposed to oil and oil
          products,

CHERRY POINT — Adult herring are spawning in
what looks like a good habitat.

                    NARRATOR (V.O.)
          . . . and the Cherry Point
          herring is perhaps the poster
          child for this effect.

RESUME — Felleman interview.

                    FRED FELLEMAN
          In fact, just after that
          refinery opened, there was a
          massive oil spill.

PHOTO — of an oil tanker moored at the
refinery dock. It has an oil spill boom
around it.

                    FRED FELLEMAN (O.S.)
          And it was during the spawn,
          and I think that was sort
          of like, uh, the tip of the
          iceberg of when things started
          going really bad for Cherry
          Point herring.

RESUME — travelling past the second oil
tanker at the Cherry Point shipping terminal.

                    KURT STICK (O.S.)
          We've observed . . .

EXT. PUGET SOUND SHORELINE — DAY

Establish. Interview. Stick is sitting on the
shore near the shipping terminal.

**LOWER THIRD: Kurt Stick, Biologist, Department
of Fish and Wildlife.**

                    KURT STICK
          . . . an increased mortality of
          older fish. The recruitment had
          been holding up pretty well of
          younger fish.

FROM THE BOAT — Stick tosses a rake attached
to a rope into the water to sample marine
plants at the bottom of the ocean for herring
eggs.

                    KURT STICK (O.S.)
          So personally, I don't think
          it's necessarily a water
          quality issue.

STICK AND A HELPER — are in the Zodiac. They
are dragging the rake across the bottom of
the ocean.

                    KURT STICK (O.S.)
          I think it's a combination of
          stressors,

UNDERWATER — The rake pulls through and grabs some algae.

AT THE SURFACE — The rake emerges with algae but no eggs.

> KURT STICK (O.S.)
> . . . natural stressors and
> man-caused, probably indirectly
> with overall environmental
> health.

WIDER — The two scientists cruise by in the Zodiac.

> JEFF MARLIAVE (O.S.)
> A lot of people would . . .

INT. MARLIAVE'S OFFICE — DAY

Establish. Interview.

LOWER THIRD: Dr. Jeff Marliave, VP Marine Science, Vancouver Aquarium.

> JEFF MARLIAVE
> . . . like to point their
> fingers at supertankers and
> say it's all their fault, but
> there's so many other things
> that could be happening
> because . . .

FROM THE SHORE — The Cherry Point terminal fills the horizon.

                    JEFF MARLIAVE (O.S.)
          . . . we've had different
          weather climate trends.

UNDERWATER — A few herring mingle, looking
for food.

                    JEFF MARLIAVE (O.S.)
          And so it's not the 97 percent
          demise of the population,

MORE HERRING — They are closer now and
feeding on zooplankton.

                    JEFF MARLIAVE (O.S.)
          . . . it's the 97 percent
          decline of the population . . .

RESUME — the terminal with the oil tanker.

                    JEFF MARLIAVE (O.S.)
          . . . at that historically
          important spawning area.

FROM UNDER A BOAT — A school of herring
navigate the waves.

                    KURT STICK (O.S.)
          It could be anywhere else in
          their range where they go. We
          think that most of these fish
          head out to . . .

RESUME — Stick interview.

                    KURT STICK
          . . . the west coast of
          Washington State and Vancouver
          Island.

AND THE SCHOOL — takes off, heading out to
sea.

RESUME — Felleman interview.

                    FRED FELLEMAN
          I was driving back from DC with
          my car and a U-Haul and . . .

PHOTO — of the *Nestucca* barge.

                    FRED FELLEMAN (O.S.)
          . . . the *Nestucca* oil spill
          happened.

PHOTO — of the *Nestucca* floating out at sea.

                    FRED FELLEMAN (O.S.)
          The barge disconnected from its
          tow and the tug was trying to
          reconnect and damaged itself,

PHOTO — of a sandy beach covered in an oily
sheen.

PHOTO — of the southwest coast of Vancouver
Island. A dead bird is on a beach covered in
oil.

> FRED FELLEMAN (O.S.)
> . . . and that oil blew up
> against the west coast of
> Vancouver Island. And in fact,
> it was, like, uh, there was a
> lighthouse keeper that saw a
> slick approaching and was so
> appalled . . .

RESUME — Felleman interview.

> FRED FELLEMAN
> . . . that nobody from the
> States bothered to tell the
> Canadians on the west coast
> what's going on.

PHOTO — of a worker picking up beach debris
covered in oil.

> FRED FELLEMAN (O.S.)
> And that started the beginning
> of this BC-Washington oil spill
> task force.

INT. PUGET SOUND — DAY

Cherry Point herring swim slowly through the
sound.

> NARRATOR (V.O.)
> Could it be that the Cherry
> Point herring were simply
> unlucky and . . .

CLOSER — The herring are moving rather slowly
and don't look healthy.

> NARRATOR (V.O.)
> . . . that oil spills like the
> *Nestucca* have impacted them at
> different parts of their life
> cycle and caused devastating
> and lasting effects?

THE HERRING — change direction a couple
times. They seem confused.

GRAPH — of the Cherry Point herring over
time. The spawning biomass drops dramatically
from 1973 to 2018.

> JEFF MARLIAVE (O.S.)
> It would seem that the Cherry
> Point stock went down at Cherry
> Point,

RESUME — Marliave interview.

> JEFF MARLIAVE
> . . . but if you're seeing fish
> at such late dates in False
> Creek,

INT. FALSE CREEK OCEAN — DAY

A large school of herring are swimming in
False Creek in April.

> JEFF MARLIAVE (O.S.)
> . . . I would bet you a cookie
> that they are genetically . . .

TIGHTER — The large school of spawners hover around another float at Fisherman's Wharf.

> JEFF MARLIAVE (O.S.)
> . . . Cherry Point fish that
> are doing better in False
> Creek than they've been doing
> recently at Cherry Point.

PHOTO — of oil spreading across upper Howe Sound from the *Westwood Anette* oil spill.

PHOTO — of the oil in the water.

SUPERSCRIPT: Squamish Estuary, August 4th, 2006.

> JOHN BUCHANAN (O.S.)
> It was a cargo vessel that was
> leaving port here in Squamish.

RESUME — Buchanan interview.

> JOHN BUCHANAN
> That was the 2006 estuary oil
> spill with the *Westwood Anette*.

EXT. SQUAMISH ESTUARY — DAY

Wind whips the sea grasses of the west side of the estuary.

PHOTO — of the sharp metal bracket that
punctured the side of the ship.

>   JOHN BUCHANAN (O.S.)
>   There were high winds and it
>   came back into port and it hit
>   hard on, uh, a couple of sharp
>   pieces of steel . . .

PHOTO — of bunker oil leaking down the beach
next to the Squamish Terminals at low tide.

PHOTO — of bunker oil spread across the
beach.

>   JOHN BUCHANAN (O.S.)
>   . . . and, uh, these two holes
>   punctured its stay tank and it
>   started spilling bunker sea oil
>   into the estuary.

PHOTO — of the oil-soaked margin between the
beach and the sea grasses.

RESUME — Buchanan interview.

>   JOHN BUCHANAN (O.S.)
>   And I showed up and I had a
>   look, and I was just appalled
>   at what I saw.

PHOTO — An oil spill response team is working
on a beach near the Terminals. The *Westwood
Anette* is still at the dock.

Oil spills occur frequently on the BC coast, but only the big ones make the news. This oil spill occurred at the Squamish Terminals on August 4, 2006. Heavy bunker oil covered a good portion of the Squamish Estuary shoreline. (Photo: John Buchanan)

Marinas are located in just about every sheltered bay on BC's south coast, and oil contamination occurs wherever there are boats. This constant contamination can impact local herring populations when they lay eggs on pilings and other surfaces at and around marinas. (Photo: Scott Renyard)

BUNKER OIL — flows like a small creek down
the beach as the tide retreats.

>   JOHN BUCHANAN (O.S.)
>   More about what I didn't see,
>   and I didn't see anybody on the
>   beach that Saturday morning.

PHOTO (DRONE) — looking down at the west side
of the Squamish Estuary. Oil has coated a
small drainage creek.

PHOTO — of the margin of the whole west side
of the estuary covered in oil.

PHOTO — of the oil on the beach facing the
east side of the estuary.

PHOTO — of a drainage ditch in the estuary
that is black with oil.

PHOTO — of sea grasses that should be green
but look burnt. But they aren't burnt.
They're covered with oil.

PHOTO (DRONE) — looking down on a large ship
with oil leaking out of it.

ANIMATED GRAPHIC — Locations, dates and
types of oil spills show over a photo of a
large ocean spill that looks like a rainbow-
coloured mosaic.

> NARRATOR (V.O.)
> The Squamish spill was just one
> of over 100 large oil spills
> that have occurred on the
> Pacific coast over the last
> eight years.

EXT. SQUAMISH HARBOUR MARINA — DAY

The marina is home to dozens of boats.

> NARRATOR (V.O.)
> And there are over 200 marinas
> and . . .

EXT. STEVESTON MARINA — DUSK

Rows and rows of boats.

EXT. WEST VANCOUVER MARINA — DAY

Hundreds of boats fill the frame.

> NARRATOR (V.O.)
> . . . thousands of boats moored
> in just about every sheltered
> bay of the southern part of
> British Columbia.

PHOTO — of a marina packed with sailboats.

PHOTO — of a large marina with several rows
of boats.

PHOTO — of a marina jammed with fishing vessels.

SEVERAL — larger fishing vessels moored at Steveston Harbour, BC.

LOOKING DOWN — The rainbow colours of an oil slick spread across the water.

> NARRATOR (V.O.)
> This results in a lot of oil finding its way into the water.

ANOTHER OIL RAINBOW — stains the surface of the water.

EXT. FALSE CREEK — DAY

Hundreds of boats are moored here.

> NARRATOR (V.O.)
> So it's possible that there is enough oil contamination in these areas . . .

UNDERWATER — A school of herring swim slowly by.

> NARRATOR (V.O.)
> . . . to cause a significant decline in herring populations.

SMALL HERRING — swim frantically through deep water.

EXT. WEST COAST VISTA (DRONE) — DAY

A spectacular view of BC's north coast. No
boats, no marinas.

A FOGGY CHANNEL — with sunlight leaking
through the clouds. The channel is devoid of
human activity.

>   NARRATOR (V.O.)
>   The problem with this theory is
>   that herring declines go beyond
>   the south coast to the . . .

A ROCKY SHORELINE — Waves crash against the
rocks. There are no marinas or boats as far
as the eye can see.

>   NARRATOR (V.O.)
>   . . . north coast, where
>   marinas and shoreline
>   developments are sparse and oil
>   spills less frequent.

A FLOCK OF GULLS — burst off a shoreline.

MORE BIRDS — dive into water filled with
herring milt. There is no other sign of human
activity across the channel.

SOME GULLS — grab herring swimming in the
shallows.

> NARRATOR (V.O.)
> So has there been enough oil
> contamination coast-wide to
> have caused a lasting decline
> in herring populations?

OVER BLACK

We hear the sound of a helicopter.

EXT. PRINCE WILLIAM SOUND (ARCHIVE) — DAY

The helicopter is flying low over a badly
oiled beach. The oily beach seems to go on
forever.

> NARRATOR (V.O.)
> The opportunity to answer this
> question came from the most
> devastating . . .

FROM A HELICOPTER — The *Exxon Valdez* is
leaking oil and creating a huge oil sheen on
the ocean.

SUPERSCRIPT: March 24, 1989.

> NARRATOR (V.O.)
> . . . oil spill that has ever
> occurred on the Pacific coast.

PANNING AROUND — the ship and a small boat
gives us a sense of scale. This is a very
large ship.

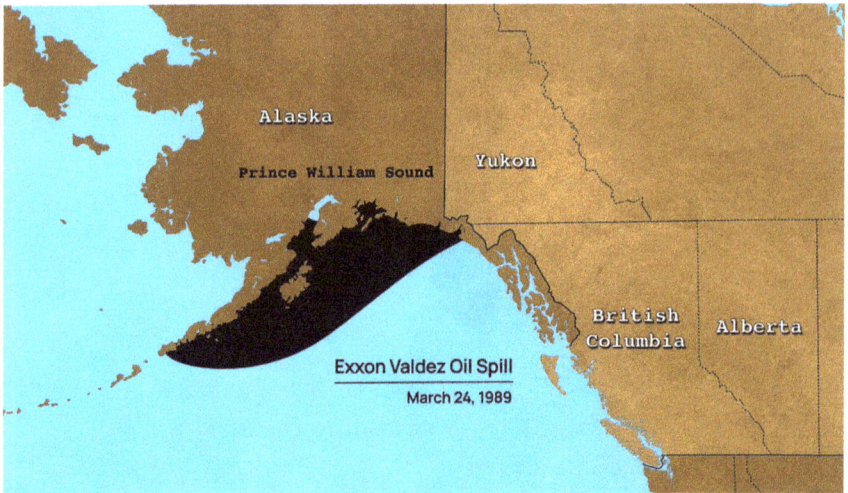

The *Exxon Valdez* oil spill on March 24, 1989, covered over 2,100 km of shoreline with heavy bunker oil, killing birds, mammals and fish. The spill happened right when herring were expected to spawn in the intertidal zone, which would have impacted herring over a vast area. (Map: Juggernaut Pictures Inc.)

NARRATOR (V.O.)
The *Exxon Valdez* is still the
second-largest oil spill in US
history,

AT A SHORELINE — The oil spill has landed on
the shore and fills our view.

NARRATOR (V.O.)
. . . and it covered 2,100
kilometres of coastline.

MOVING AROUND — the back of the ship.

RICK STEINER (O.S.)
The *Exxon Valdez* was a loaded
single-hull supertanker,

INT. RICK STEINER OFFICE — DAY

Establish. Interview.

**LOWER THIRD: Richard Steiner, Retired
Professor, University of Alaska.**

RICK STEINER
. . . loaded with Alaska North
Slope crude oil. Left the
Valdez Marine Terminal. Slammed
full speed into Bligh Reef in
Prince William Sound.

CLOSE — on the ship's name.

OFF THE EDGE — of the ship. Thick oil covers the water.

> RICK STEINER (O.S.)
> Ruptured 8 of its 11 oil cargo tanks and spilled a minimum of 11 million gallons,

PUSHING IN — The oil is a terrible sight.

> RICK STEINER (O.S.)
> . . . and actually we think it was probably two to three times that amount came out of the tanker.

ANOTHER SHORELINE — covered in thick, black oil.

> RICK STEINER (O.S.)
> The oil flowed southwest through Prince William Sound. It oiled over 10,000 square miles of Alaska's coastal ocean.

RESUME — Steiner interview.

> RICK STEINER
> Millions of innocent animals were killed that first year alone.

A YOUNG SEAL — dead and coated with oil, is
pulled out of the shallow water by an oil
spill worker.

> RICK STEINER (O.S.)
> Thousands of marine
> mammals . . .

ANOTHER WORKER — carries a cormorant
completely covered in oil. It tries to nip
its rescuer.

> RICK STEINER (O.S.)
> . . . and hundreds of thousands
> of seabirds.

A CLEAN-UP WORKER — pokes at the oil on the
rocks with a stick.

CLOSER — He stirs the black oil soup with the
stick. It's like tar.

> RICK STEINER (O.S.)
> Oil ended up on beaches. It was
> a horrible, horrible sight to
> see.

MAP — of the 2,100-mile range of the *Exxon
Valdez* oil spill.

> DOUG HAY (O.S.)
> In the Prince William Sound
> area, oil was spilled in 1989
> in areas where herring did
> spawn,

RESUME — Hay interview.

LOWER THIRD: Dr. Douglas Hay, Scientist Emeritus, Fisheries and Oceans Canada.

>   DOUG HAY
> . . . and there's still
> arguments about the degree of
> impact of oil on the eggs at
> that time.

INT. EGG STUDY LABORATORY (RE-ENACTMENT) — DAY

Herring eggs are being kept alive in a series of containers to see the effects of oil and oil products on developing herring embryos.

THROUGH A MICROSCOPE — An eyed egg looks back at us.

>   NARRATOR (V.O.)
> Egg studies immediately
> following the disaster
> found herring embryos with
> abnormalities consistent with
> oil contamination, which
> include . . .

MACRO — of a herring larva with a swollen yolk sac.

> NARRATOR (V.O.)
> . . . swollen yolk sacs,

MACRO — of a herring larva with its heart beating irregularly.

> NARRATOR (V.O.)
> . . . heart defects,

MACRO — of a herring larva with a twisted spine.

> NARRATOR (V.O.)
> . . . spinal deformities . . .

MACRO — of a herring larva with a jaw missing.

> NARRATOR (V.O.)
> . . . and jaw malformations.

INT. LIVING ROOM — DAY

Establish. Interview.

LOWER THIRD: Dr. Gary Marty, Research Associate, School of Veterinary Medicine, University of California, Davis.

> GARY MARTY
> Well, I first got involved in Prince William Sound, Alaska, after the 1989 *Exxon Valdez* oil spill.

GRAPH — of the Prince William Sound herring biomass beginning in 1974. The biomass grows until just before 1989. A red dot appears just at the time of the oil spill.

> GARY MARTY (O.S.)
> This was confounded with Pacific herring in the fact that their population was at historic highs when the oil spill occurred in 1989,

FROM ABOVE — The crew of a barge equipped with a crane and high-pressure hose are cleaning the rocks.

WORKERS WITH HIGH-PRESSURE HOSES — spray water on the beach as they try to disperse the oil.

> GARY MARTY (O.S.)
> . . . but it occurred in the end of March, which is also their spawning period, so there was concern that the herring would be affected.

ON A HOSE — A worker sprays an oil-covered rock.

> GARY MARTY (O.S.)
> They closed the fishery in 1989 because of concerns of oil contamination.

INT. OCEAN (1989) — DAY

A few herring swim by rockweed with fresh
eggs on it.

                GARY MARTY (O.S.)
          So now for the next several
          years you have a population
          that's at historic highs.

RESUME GRAPH — The Prince William Sound
herring biomass crashes by 1993.

                GARY MARTY (O.S.)
          In 1993, they came back, four
          years after the spill, and the
          population had collapsed.

RESUME — Marty interview.

                GARY MARTY
          One research group thinks
          that oil was the cause of the
          decline.

RESUME (1989) — archival coverage of the
Exxon Valdez oil spill clean-up. The crew of
the barge crane are washing the beach.

                GARY MARTY (O.S.)
          I disagree with this assessment
          because the oil that we
          can detect in herring had
          disappeared within a year of
          the spill.

ON THE HIGH-PRESSURE HEAD (1989) — of a water sprayer. It is a couple of metres wide and can soak a wide area.

INT. OCEAN (1989) (RE-ENACTMENT) — DAY

Herring travel slowly along the shoreline.

> GARY MARTY (O.S.)
> Herring are able to metabolize
> the oil quite quickly, and so
> the oil is not persistent in
> the herring tissues.

A WIDER VIEW (1989) (RE-ENACTMENT) — The oil spill crew are spread out across a rocky beach and continue to wash the oil into the water.

> GARY MARTY (O.S.)
> The best evidence indicates
> that the oil spill killed less
> fish than . . .

THE OIL SPILL CREW (1989) — work over a few rocks with high-pressure hoses.

> GARY MARTY (O.S.)
> . . . the fishery would have if
> it had occurred.

FROM A WORKER'S POV (1989) — They work their hose back and forth over the oil-covered rocks.

> GARY MARTY (O.S.)
> The decline occurred three
> years later,

MORE WORKERS (1989) (RE-ENACTMENT) — stretch
out a rope with oil-absorbent pads attached
to it.

RESUME GRAPH — of herring population in
Prince William Sound. The biomass doesn't
recover.

> GARY MARTY (O.S.)
> . . . so there's no plausible
> mechanism that can explain
> how the oil one year after the
> spill killed the fish in the
> population three or four years
> after the spill.

INT. PRINCE WILLIAM SOUND OCEAN (1989) (RE-
ENACTMENT) — DAY

Herring packed tightly together look very
sick and are covered in skin lesions.

> NARRATOR (V.O.)
> And many of the spawners had
> skin lesions.

ANOTHER WORKER (1989) (RE-ENACTMENT) — picks
up an oil-covered bird. It's dead.

A DEAD SEAL (1989) — is pulled out of the
water.

Researchers initially believed that the crash in herring populations in 1993 stemmed from the 1989 *Exxon Valdez* oil spill. But research led by Dr. Gary Marty discovered that a virus called viral hemorrhagic septicemia (VHS) was the likely cause of the herring population crash. This map shows the range of VHSv on the Pacific coast in 1993. There is evidence that this virus was introduced to the region in the late 1970s in Puget Sound and moved north via the overlapping migrations of wild fish populations, which spread the virus. (Map: Juggernaut Pictures Inc.)

This herring, sampled on June 13, 2013, at Blenkinsop Bay in the Discovery Islands on the east side of Vancouver Island, appears sunburnt. This is a typical symptom of a VHS infection. The fish is bleeding into its flesh. (Photo: Jody Eriksson)

This herring, also sampled at Blenkinsop Bay on June 13, 2013, is bleeding from the creases of its fins. This is another symptom of a VHS infection. (Photo: Jody Eriksson)

This herring, sampled at the Sointula, BC, public dock on August 10, 2013, is bleeding around the mouth and base of its fins. This type of bleeding is a common symptom of a VHS infection. (Photo: Alexandra Morton)

                    NARRATOR (V.O.)
          And because there was no large
          single herring die-off event
          like that seen in birds and
          other animals,

TWO WORKERS — place an otter in a bag.

ANOTHER WORKER — tosses a dead bird down the
beach.

                    NARRATOR (V.O.)
          . . . the evidence seemed to
          point away from the oil spill
          and towards something else.

DOCUMENT — Marty et al. (1998). Viral
hemorrhagic septicemia virus, *Ichthyophonus
hoferi*, and other causes of morbidity in
Pacific herring *Clupea pallasi*. The words
"viral hemorrhagic septicemia virus (VHSv)
was isolated" lift off the page.

                    GARY MARTY (O.S.)
          They didn't have any mechanism
          in place at that time to study
          the collapse at a population
          level, but viral hemorrhagic
          septicemia virus was isolated
          from the fish.

PHOTO — of five gizzard shad covered in
bleeding lesions.

PHOTO — of a giant Fraser River sturgeon with
a red "burnt" belly.

> GARY MARTY (O.S.)
> It's a member of the Nora
> rhabdovirus.

PHOTO — of walleye covered in red bulbous and
bleeding lesions.

> GARY MARTY (O.S.)
> It's kind of a complex name.

PHOTO — of a pink salmon from Shovelnose
Creek in the Squamish watershed.

> GARY MARTY (O.S.)
> It's a virus that affects many
> different species of fish.

PHOTO — of three more gizzard shad with VHSv
lesions.

PHOTO — of a gizzard shad with VHSv lesions.

> GARY MARTY (O.S.)
> It turned into the most
> comprehensive study of disease
> on a wild fish population.

RESUME — Marty interview.

> GARY MARTY
> In the end, we had 13 years of
> data. And we were able to get a
> pretty good idea of the role of
> disease in population decline
> of these herring.

PHOTO — of a herring with a bleeding mouth
and lesions on its belly.

PHOTO — of a herring with a bleeding mouth
and a blood spot in its eye.

PHOTO — of a herring with blood leaking from
the creases of its fins.

PHOTO — of a herring with a bleeding mouth
and a blood spot in its eye.

EXT. GERMAN RAINBOW TROUT FARM (1938) (RE-
ENACTMENT) — DAY

A fish farmer dumps some rainbow trout from a
hand net into a pen.

> NARRATOR (V.O.)
> Viral hemorrhagic septicemia
> was first discovered in Germany
> in 1938.

EXT. DANISH RAINBOW TROUT FARM (1962) (RE-
ENACTMENT) — DAY

The pen is full of large mature rainbow trout.

>   NARRATOR (V.O.)
>   And was later found in a Danish
>   fish farm raising rainbow trout
>   in 1962.

THE TROUT — are impressively large.

>   NARRATOR (V.O.)
>   Shortly after that, Norwegian
>   fish farmers developed open net
>   pens and began growing rainbow
>   trout and Atlantic salmon in
>   the ocean.

INT. NORWEGIAN FISH FARM (1962) (RE-
ENACTMENT) — DAY

The large open net pen is full of thousands of Atlantic salmon.

>   NARRATOR (V.O.)
>   Other countries like Japan,
>   Scotland and Canada jumped into
>   the aquaculture game . . .

A CONTAINER — of salmon eggs is scooped out of a hatchery.

CLOSER — on the eggs.

This fish farm located in Okisollo Channel in the Discovery Islands is a typical example of the farms on the BC coast. BC fish farms, which usually raise Atlantic salmon, are made up of a number of square pens attached to each other. (Photo: Tavish Campbell)

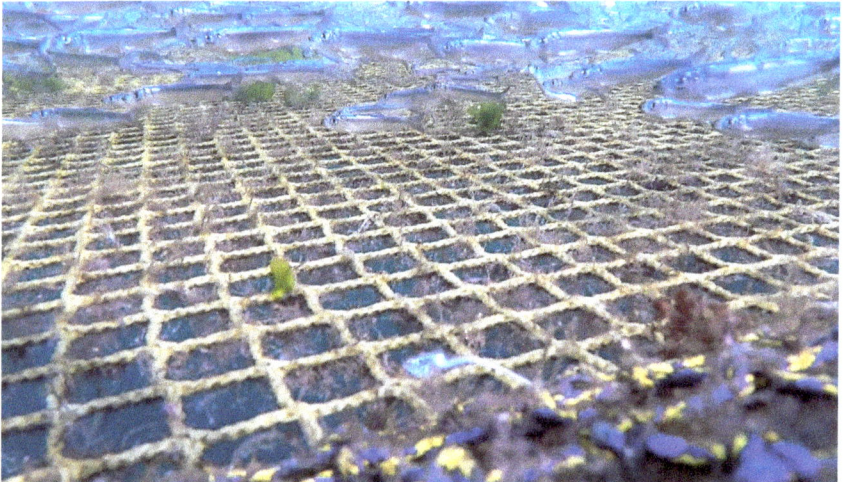

Pacific herring mingle in and around open net pen fish farms. They are attracted to them because they are able to graze on the finer particles of the fish farm food that come off the fish farm pellets. Smaller herring that enter the pens can become prey for the farmed Atlantic salmon and essentially subsidize the food costs for the fish farm operation. This close contact between wild and farm fish increases transmission of diseases between fish populations. (Photo: Alexandra Morton)

Shiner perch (*Cymatogaster aggregata*) are frequently found in and around open net pen fish farms. (Photo: Alexandra Morton)

Juvenile herring swimming around the public docks at Campbell River, BC, on July 25, 2014, were covered in sea lice. Sea lice can carry viruses and bacteria from one fish host to another. Open net pen fish farms are located near Campbell River and provide ideal conditions for sea lice populations to expand by providing many hosts in a small area. Research on sea lice at fish farms suggests this is the reason these herring have heavy sea lice loads. (Photo: Scott Renyard)

> NARRATOR (V.O.)
> . . . and began trading live
> fish products like Atlantic
> salmon eggs, which helped to
> spread the VHS virus across the
> globe.

AT A DOCK — Hundreds of small fish are
floating on the surface of the water, dead.

> NARRATOR (V.O.)
> A recent count found the virus
> has infected more than 82
> species of fish.

ON A BEACH — Hundreds of fish are dead. There
are several species in the mix.

A PILE OF DEAD HERRING — They all have bloody
heads and eyes.

> GARY MARTY (O.S.)
> The type that we have in the
> eastern Pacific is different
> from the type that was found in
> the Great Lakes, and it's also
> different from the type that's
> found in Europe.

RESUME — Marty interview.

> GARY MARTY
> The type here in the eastern
> Pacific tends . . .

HUNDREDS OF HERRING — lie in shallow water.
Most are dead, but a few of them are still
alive.

>   GARY MARTY (O.S.)
>   . . . to occur in herring in
>   quite large numbers and then
>   it can affect other species at
>   various times. If three percent
>   of the fish had ulcers during
>   their spawning time, that was
>   associated with significant
>   population decline.

INT. SALISH SEA OCEAN (1970s) (RE-ENACTMENT)
— DAY

HERRING — swimming slowly in a dispersed
group. This is not a normal herring pattern.

>   NARRATOR (V.O.)
>   Pacific herring populations
>   began to decline on Canada's
>   west coast in the late 1970s.

ON THE BOTTOM — There are no herring to be
seen here.

ANIMATED MAP — The colour of the ocean turns
red, starting in Puget Sound and spreading
north. A legend showing the years turns over
year by year from 1978 to 1989, when the red
reaches Prince William Sound in Alaska.

                    NARRATOR (V.O.)
          The Department of Fisheries
          and Oceans records reveal this
          began near the southwest corner
          of Vancouver Island and moved
          north, arriving in Alaska a
          decade later, right around the
          time of the *Exxon Valdez* oil
          spill.

INT. PORT RUPERT HARBOUR (1993) (RE-
ENACTMENT) — DAY

Dead herring cover the bottom of the ocean.

PHOTO — of two dead herring with VHSv.

                    NARRATOR (V.O.)
          And the first official
          detection of VHSv came in 1993,
          when herring with lesions were
          found dead after a diesel spill
          at Prince Rupert Harbour.

INT. UNDER A WHARF OCEAN — DAY

Herring mingle under the wharf.

CLOSER — The herring swim through the murky
green water.

> NARRATOR (V.O.)
> It's no wonder that there's
> confusion about what's
> affecting herring when there
> are different factors operating
> simultaneously in the same
> areas,

THE HERRING — are now feeding on something.

> NARRATOR (V.O.)
> . . . which may or may not
> contribute to a disease event.

EVEN CLOSER — The herring are grabbing
copepods out of the water.

EXT. BC COAST — DAY

A fish farm hugs the far side of the inlet.

> NARRATOR (V.O.)
> Since 1998, there have been
> many cases of herring die-offs
> on the British Columbia coast.

TRAVELLING — outside a pen.

CLOSER — The pens are covered in predator
nets.

>   NARRATOR (V.O.)
>   And researchers found that
>   British Columbia fish farms
>   tested positive for viral
>   hemorrhagic septicemia at a
>   population level . . .

LOOKING INTO A PEN — Large Atlantic salmon
mill around in it.

UNDERWATER — The farm fish swim in a circle
inside the pen.

>   NARRATOR (V.O.)
>   . . . and believe that fish
>   farms are acting as reservoirs
>   for the virus.

LOOKING STRAIGHT DOWN (DRONE) — A feeder is
operating at a pen and the farmed fish in the
pen are at the surface of the water.

PUSHING IN — Hundreds of large farmed salmon
mingle in the pen near the feeder.

>   NARRATOR (V.O.)
>   Some farmers claim that VHS
>   virus has always been on the
>   British Columbia coast,

PANNING ACROSS — the farm. The pens are
covered in mesh to stop predators from
getting at the farmed fish.

> NARRATOR (V.O.)
> . . . but there are plenty of
> records indicating that this is
> not the case.

LOOKING THROUGH — the netting covering a pen.
A couple of farmed fish come up for a breath
of air.

> NARRATOR (V.O.)
> And now that the virus is
> so widespread, this disease
> continues to haunt the open net
> fish farms.

TRAVELLING AROUND — the perimeter of a fish
farm. All is quiet.

EXT. FISH FARM SUPPORT SHIP — DAY

Small farm fish are being pumped into the
hold of the ship.

FROM ABOVE — The fish come out of an orange
hose in a stream of water.

> NARRATOR (V.O.)
> Canadian scientists found that
> when VHS virus-free Atlantic
> salmon smolts were released
> into open net pens,

A SCHOOL — of wild sardines dart past a farm.

> NARRATOR (V.O.)
> . . . wild sardines,

WILD PERCH — mingle in the farm with Atlantic salmon.

> NARRATOR (V.O.)
> . . . wild perch and . . .

JUVENILE HERRING — by the thousands swim just outside a fish farm pen.

> NARRATOR (V.O)
> . . . wild herring mingling
> around the farms died from VHSv
> a month later.

MORE HERRING — are inside the pen, but these ones are adults.

> NARRATOR (V.O.)
> How could that be?

LOOKING THROUGH — from the outside of a net pen. Large Atlantic salmon swim slowly inside it.

INSIDE THE NET PEN — a school of herring swim with the much larger Atlantic salmon.

> NARRATOR (V.O.)
> At the time, researchers
> concluded that the wild herring
> and . . .

A SCHOOL OF SARDINES — swim gracefully
together.

                NARRATOR (V.O.)
         . . . sardines infected the
         farmed fish, which could very
         well be true.

INSIDE A FARM — Large Atlantic salmon swim
slowly toward camera.

                NARRATOR (V.O.)
         But what also likely occurred
         is, the farmed fish altered or
         amplified the virus after it
         passed through their bodies,

TRAPPED WILD PERCH — mingle with the farmed
fish inside a pen.

TRAPPED HERRING — in the pen swim together
and try to avoid the bigger Atlantic salmon.
They suddenly split up to avoid an Atlantic
salmon going the other way.

                NARRATOR (V.O.)
         . . . then reinfected the
         herring and other wild fish
         with an altered virus that had
         lethal results.

CLOSER ON THE FARMED SALMON — They are larger
and most seem to be healthy.

> NARRATOR (V.O.)
> The delayed reaction of the
> farmed fish die-off might mean
> they were initially healthier
> than the wild fish and it
> took longer for the original
> infection to take hold.

LOOKING DOWN — into the pen. Several dead
salmon are floating on the surface of the
water.

> NARRATOR (V.O.)
> Or the wild fish changed the
> virus as it went through their
> bodies and passed it back to
> the farmed fish, which made it
> more lethal to them.

BACK TO — the orange hose pumping small
farmed fish into the farm pens.

> NARRATOR (V.O.)
> Whatever the mechanism,

IN AN OPEN NET PEN — the salmon are swimming
suspiciously slowly near the surface of the
water.

> NARRATOR (V.O.)
> . . . it's clear that the
> introduction of Atlantic salmon
> smolts into . . .

A SMALL DEAD FARMED SALMON — floats inside the pen.

ANOTHER ONE — floats belly-up.

> NARRATOR (V.O.)
> . . . an open net pen triggered
> a series of events that caused
> a die-off.

SEVERAL MORE DEAD SALMON FLOAT — in the foamy water.

ANOTHER — gasps for a breath and sinks back into the water.

INT. SALISH SEA — DAY

Juvenile herring are swimming along just under the surface.

> NARRATOR (V.O.)
> Researchers found under
> laboratory conditions,
> infectious salmon anemia virus,
> or . . .

CLOSER ON THE HERRING — They are covered in sea lice.

> NARRATOR (V.O.)
> . . . ISA, spreads quickly from
> fish to fish when sea lice are
> present.

ON OTHER HERRING — A large school are trapped in a pen but show no obvious evidence of sea lice.

> NARRATOR (V.O.)
> And when sea lice were absent, viral transmission did not occur at all.

ON A PINK SALMON JUVENILE — covered in sea lice.

> NARRATOR (V.O.)
> This means that sea lice carry pathogens from one host to another.

TIGHT ON — a fast-moving young adult sea louse, *Caligus clemensi*, swimming in a circle.

> NARRATOR (V.O.)
> And what is worse is the sea louse,

VERY CLOSE — on a sea louse chewing on the back of a pink salmon smolt.

> NARRATOR (V.O.)
> . . . *Caligus clemensi*, parasitizes many species of fish.

INT. LARGE FISH FARM — DAY

A huge school of herring are in a fish farm
pen. An Atlantic salmon takes a run at them
and they scatter.

INSIDE THE PEN — another Atlantic salmon
chases the herring along the net pen.

> NARRATOR (V.O.)
> So this means it can jump from
> an Atlantic salmon in a fish
> farm to a herring and back
> again.

EXT. FISH FARM (UNDERWATER) — DAY

Looking up to the surface, there is a variety
of fish species mingling around the outside
of the farm.

> NARRATOR (V.O.)
> And to any other species of
> fish in the area.

ANOTHER VIEW — Dozens of needlefish swim up
to the net pen.

> FADE OUT:

FADE IN:

EXT. SOINTULA MARINA — DAY

All is quiet at the marina. Boats are moored, and there is a slight breeze. No one is around.

SUPERSCRIPT: August 10, 2013.

> NARRATOR (V.O.)
> In the summer of 2013, there were large schools of herring feeding off the docks at Sointula, British Columbia.

INT. SOINTULA MARINA OCEAN — MOMENTS LATER

Lots of herring are swimming and feeding in and around the docks.

CLOSER — The herring have sea lice on them.

> NARRATOR (V.O.)
> At first, it was exciting to see so many herring. But upon taking a closer look, nearly all the fish had lots of sea lice on them. And most . . .

EXT. PHOTARIUM — DAY

A herring is bleeding from its belly and face.

> NARRATOR (V.O.)
> . . . of them had skin lesions
> and were bleeding.

INT. CAMPBELL RIVER MARINA OCEAN (2014) — DAY

Young adult herring are busy feeding, just like they were in Sointula, BC, the previous year.

CLOSE — The young herring have lots of sea lice on them.

> NARRATOR (V.O.)
> The following summer, herring
> swimming at the Campbell River
> public docks also had very
> heavy sea lice infestations.

A LARGER SCHOOL — swim close. They are also covered in sea lice.

> NARRATOR (V.O.)
> These fish were in worse
> shape than the herring seen in
> Sointula the year before.

EXT. FALSE CREEK OCEAN (2019) — DAY

Small herring swim by. They have lots of sea lice on them.

CLOSER — and we can see the sea lice more clearly.

                    NARRATOR (V.O.)
          And a school of immature
          herring swimming in False Creek
          were covered with sea lice,

NEARBY — More juvenile herring with sea lice
on them.

                    NARRATOR (V.O.)
          . . . some with open lesions
          and mucus trailing from their
          vents, indicating infection.

EXT. FISHERMAN'S WHARF OCEAN (2019) — DAY

Herring spawners have what look like nicks on
their side. These marks are a telltale sign
of sea lice.

                    NARRATOR (V.O.)
          Even the adult spawners feeding
          in False Creek in 2019 had lots
          of small scars and immature sea
          lice on them.

MAP — of BC's south coast and many locations
where researchers have found high sea lice
loads on juvenile salmon and herring. Then
the whole Inside Passage goes red.

                    NARRATOR (V.O.)
          Spanning more than two decades,
          researchers have documented
          high sea lice loads on herring
                    (MORE)

          NARRATOR (V.O.) (CONT'D)
and juvenile salmon throughout
the Inside Passage. This
phenomenon spans at least 400
kilometres.

A WIDER MAP — shows the Pacific herring
range. It extends right across the Pacific
Ocean and represents the potential range
within which these infections can be found
in Pacific herring.

          NARRATOR (V.O.)
And likely extends throughout
the Pacific herring range.

A RESEARCHER'S NET — filled with herring and
juvenile sockeye salmon.

          NARRATOR (V.O.)
And even more alarming, many
of the samples taken by
researchers found sockeye and
herring juveniles swimming
in close proximity with each
other.

INT. OCEAN — DAY

Sockeye salmon juveniles navigate ocean
currents and waves in shallow water.

          NARRATOR (V.O.)
The widespread loss in sockeye
productivity corresponds . . .

INT. SQUAMISH TERMINALS OCEAN — DAY

There are only a few herring zipping around.

>   NARRATOR (V.O.)
>   . . . in time and scope with
>   the loss of productivity in
>   herring.

HERRING — with heavy sea lice loads are
swimming just under the surface of the water.

>   NARRATOR (V.O.)
>   So it's likely that the
>   persistent sea lice plague
>   is playing an active and key
>   role in the struggles of both
>   species.

CLOSER — The herring swimming by are loaded
with sea lice.

ANOTHER SHOT — shows an even denser school
of juvenile herring with sea lice. This
indicates that it is very easy for sea lice
to jump from one herring to another.

>   NARRATOR (V.O.)
>   The two species are
>   experiencing catastrophic
>   challenges with disease
>   outbreaks.

EXT. OPEN NET PEN FISH FARM — DAY

Large support boats are moored at this farm.

ONE OF THE SHIPS — is pumping smolts into the pens.

> NARRATOR (V.O.)
> And all of this is happening
> in an area with the highest
> concentration of open net pen
> fish farms on North America's
> Pacific coast.

ON THE PIPE — Water and fish pass from the ship to the farm.

> NARRATOR (V.O.)
> But unfortunately, there is
> even more bad news.

INT. FISH FARM OCEAN — NIGHT

Just outside the pen, hundreds of herring mingle around the outside of the farm.

RESUME — Marty interview.

LOWER THIRD: Dr. Gary Marty — Research Associate, University of California, Davis.

                    GARY MARTY
          The second thing was a
          primitive organism called
          Ichthyophonus hoferi. This is
          an organism that is somewhere
          between the fungi and animals,

LOTS OF HERRING — lie dead at the bottom of
the ocean.

                    GARY MARTY (O.S.)
          . . . and it's known to kill
          large numbers of herring
          and other related species
          worldwide.

DOCUMENT — Gozlan et al.(2014). Current
ecological understanding of fungal-like
pathogens of fish: What lies beneath?

PHRASES PEEL OFF THE PAGE — "increasing fish
movement around the world" and then "for
farming."

PAN DOWN — as more phrases peel off:
"infectious agents," "may have benefited from
recent increase" and then "in global trade."

                    NARRATOR (V.O.)
          This fungal-like parasite is
          on the rise globally and with
          "increasing fish movement"
          around the world for farming,
          "infectious agents," as one
                    (MORE)

                    NARRATOR (V.O.) (CONT'D)
          author wrote, may have
          "benefited" from a recent
          increase in "global trade."

PACIFIC HERRING — swim right at us over a bed
of rockweed, then dodge to the left.

OTHER HERRING — swim through a thick cloud of
milt.

                    NARRATOR (V.O.)
          Researchers have found that
          this parasite often infects
          both Atlantic and Pacific
          herring at a 25 to 30 percent
          rate, with spikes in older
          herring populations reaching as
          high as 70 percent.

EXT. DISCOVERY ISLANDS FISH FARM (DRONE) —
DAY

An eight-pen farm dominates the landscape.

                    NARRATOR (V.O.)
          The life cycle of *Ichthyophonus
          hoferi*, or ICH, is not
          completely understood,

A SPEEDBOAT — roars past the fish farm.

INSIDE A FARM — herring and salmon swim close
to each other.

                    NARRATOR (V.O.)
          . . . but what is known is that
          it requires more than one host
          to complete its life cycle.

A SMALL SCHOOL — of herring scoot along a net
in a fish farm. Lots of Atlantic salmon lurk
in the background.

                    NARRATOR (V.O.)
          Under normal circumstances,
          ICH is believed to pass from
          pelagic migratory herring to
          demersal non-migratory species
          and back again,

SEA LICE-LADEN — herring swim erratically
just under the surface of the water.

                    NARRATOR (V.O.)
          . . . possibly using
          invertebrates, such as sea
          lice, or water-borne cells as a
          conduit.

CLOSER ON — herring with heavy sea lice
loads.

IN THE SHALLOWS — a herring lies dying in the
seaweed.

                    NARRATOR (V.O.)
          But the rise in the rate of ICH
          infection over recent decades
          suggests that something has
          changed.

EXT. FISH FARM SKY (DRONE) — DAY

Looking down at a fish farm pen crowded with
farmed fish.

                    NARRATOR (V.O.)
          So the question is, have fish
          farms provided a new way for
          this parasite to complete its
          life cycle . . .

THROUGH THE NET — Farmed salmon swim slowly
around the pen.

                    NARRATOR (V.O.)
          . . . by turning migratory
          fish like Atlantic salmon into
          abundant stationary hosts . . .

FROM ALONG — a net inside a pen. A lot of
herring are inside the farm.

                    NARRATOR (V.O.)
          . . . that are like the
          demersal hosts in nature?

TWO PINK SALMON JUVENILES — loaded with sea
lice.

> NARRATOR (V.O.)
> And by providing millions of
> sea lice,

CLOSE — The pink smolt has a sea louse and
dark spots on its side.

> NARRATOR (V.O.)
> . . . is the ICH life cycle now
> much easier to complete?

A SCHOOL OF HERRING — swim slowly through
thick green water.

> NARRATOR (V.O.)
> And perhaps it's more than
> curious that the rate of ICH
> infection virtually matches the
> rate of yet another infection.

THE SCHOOL — swim through a beam of light.

> GIDEON MORDECAI (O.S.)
> ENV, or erythrocytic necrosis
> virus. It's a DNA virus that
> infects fish.

EXT. MORDECAI BACKYARD — DAY

Establish. Interview. Dr. Gideon Mordecai
specializes in the ecology of viruses and
discovered 15 new viruses in salmon.

During a herring ponding experiment conducted by the DFO, the ENV infection rate increased from 3% to 70% after a few months. This photo shows an example of herring kept in a pond that are clearly sick and covered in lesions. (Photo: Juggernaut Pictures Archives)

Two Pacific herring with their gill plates removed. The top herring was anemic and suffering from ENV. The pale gill is a clinical sign of ENV. The bottom herring was normal. (Photo: Paul Herschberger, US Geological Survey)

LOWER THIRD — Dr. Gideon Mordecai, Research Associate, Institute of Oceans and Fisheries, UBC.

> GIDEON MORDECAI
> It actually targets the blood
> cells of, of fish and we know
> it infects a range of different
> species,

A LARGE SCHOOL — of adult sockeye salmon in a deep lake.

> GIDEON MORDECAI (O.S.)
> . . . including all five
> species of Pacific salmon. And
> also . . .

HERRING — swim quickly over a rockweed patch.

> GIDEON MORDECAI (O.S.)
> . . . herring and . . .

A POLLOCK — rests near the bottom of the ocean.

AND LASTLY — a rockfish disappears behind a rock.

> GIDEON MORDECAI (O.S.)
> . . . pollock and a range of
> species. And it's found all
> around the Northern Hemisphere.

MONTAGE: Three shots of herring feeding on
copepods.

> GIDEON MORDECAI (O.S.)
> It's probably best studied in
> herring, and in herring we know
> it's associated with a blood
> disease. The virus infects the
> red blood cells of the fish,

PHOTO — of two dead herring with their gill
plates removed. The top herring has pale
gills and the bottom one has normal red
gills. The word "anemic" fades up under the
first fish, followed by "normal" under the
second fish.

> GIDEON MORDECAI (O.S.)
> . . . and this leads to
> anemia.

PUSHING IN — on the fish with the pale gills.

> GIDEON MORDECAI (O.S.)
> So one of the first things
> you'll note in a fish which is
> infected is the gills are pale,

RESUME — Mordecai interview.

> GIDEON MORDECAI
> . . . um, and this is like one
> of the most obvious signs that
> there's something going on with
> the blood of that fish.

BAR GRAPH — shows ENV infection rates in several species of fish. (Source: Pagowksi et al., 2019.)

But the rate in this case isn't as important as the fact that the DNA of the ENV infecting chinook salmon and the ENV infecting herring are 99 percent identical, meaning there is a strong link between the species.

PUSHING IN — on the table. The chinook and herring bars turn red.

> NARRATOR (V.O.)
> What the scientists also noticed was that the ENV infecting herring was 99 percent identical to the ENV infecting chinook salmon.

UNDERWATER — A school of chinook salmon swim toward us.

> NARRATOR (V.O.)
> And this "high genetic similarity" between ENV obtained from salmon and herring . . .

CLOSER — Large chinook salmon cruise through the pool.

                    NARRATOR (V.O.)
          . . . also suggests that since
          chinook salmon prey on herring,
          they likely acquire the
          virus . . .

SEA LICE-INFECTED — herring feed around some
pilings.

                    NARRATOR (V.O.)
          . . . by eating an infected
          herring.

MORE — herring infected with sea lice.

                    NARRATOR (V.O.)
          Or they are bitten by a sea
          louse that was feeding on an
          infected herring.

INT. FISHERIES DEPARTMENT POND (1998) (RE-
ENACTMENT) — DAY

Herring are mingling in a pond.

UNDERWATER — The herring are spawning on the
mesh around the pond.

                    NARRATOR (V.O.)
          In 1988, during a ponding
          experiment, captive herring
          started with an initial three
          percent ENV infection rate,

CLOSER — The herring continue to spawn on the sides of the pen.

> NARRATOR (V.O.)
> . . . and after five months, 70 percent were infected.

THE HERRING — swim into another section of the pen.

> NARRATOR (V.O.)
> This experiment likely sent shivers down the spines of fisheries managers dealing with the growing aquaculture industry.

THE SCHOOL — swims together to the far side of the pond.

> NARRATOR (V.O.)
> It showed that confining fish results in dramatic increases in the rate of disease.

NOW — the herring swim under an overhead walkway.

THE HERRING — swim along the floor of the pen.

FADE TO:

INT. WEST COAST STREAM — DAY

A pair of coho spawners are getting ready to
spawn.

A COHO FEMALE — turns on her side to dig a
redd.

> GIDEON MORDECAI (O.S.)
> I was on a viral discovery
> project, and my job was to try
> to find new viruses in salmon.

A MALE COHO — swims through the silt stirred
up by a female.

AND A COHO — with a patch of fungus on its
tail hides under a bank.

> GIDEON MORDECAI (O.S.)
> What we quickly learned, and
> what we're learning, is we know
> very, very little about the
> diversity of viruses that are
> in fish.

RESUME — Mordecai interview.

> GIDEON MORDECAI
> I think in the few years I was
> working on viral discovery we
> discovered 15 new or emerging
> viruses in salmon.

BACK TO HERRING — feeding.

>GIDEON MORDECAI
>I imagine if you worked on
>herring and did the same study,
>you'd find something similar.

MORE HERRING — They look sick.

A FEW SICK — herring are swimming slowly.

CLOSER — Sea lice can even be seen on a
herring's head.

>NARRATOR (V.O.)
>There are two things about
>herring that make them
>especially dangerous in the
>marine food web if they become
>sick.

TWO SHOTS OF A HERRING BALL — It is a frenzy
of fish jammed tightly together.

>NARRATOR (V.O.)
>First, because herring
>instinctively swim in tight
>schools to fend off predators,
>it's easy for disease to spread
>through the population.

HERRING — with lots of sea lice on them swim
slowly.

                    NARRATOR (V.O.)
          And the speed with which the
          disease spreads is ramped up
          even more when the herring are
          exposed to unnaturally high
          levels of sea lice.

IN THE SHALLOWS — two herring schools merge
into one.

                    NARRATOR (V.O.)
          And because there is such a
          high rate of intermingling
          between herring stocks, a
          disease has few barriers and
          can spread right across the
          coast.

EXT. BRITISH COLUMBIA COAST — DAY

The ocean stretches out alongside the
mountains.

INT. FALSE CREEK OCEAN — DAY

Young translucent herring are a wriggling
mass of activity.

                    NARRATOR (V.O.)
          Second, since herring play a
          key role in the food web at all
          stages of their life,

TIGHTER — Young translucent herring mingle in a bay near docks.

A MASSIVE SCHOOL — of herring juveniles shows how easy it might be for a pathogen in this group to spread to fish that eat the herring.

                    NARRATOR (V.O.)
          . . . virulent pathogens like
          VHS have multiple opportunities
          to spread to other species of
          fish.

RESUME ENV BAR GRAPH — This time the herring and chinook bars fade slightly and the stickleback bar goes red. "40% Infection Rate" fades up.

                    NARRATOR (V.O.)
          In the Mordecai study,
          sticklebacks tested positive
          for ENV 40 percent of the time.

UNDER A WHARF — thousands of sticklebacks form an impressive school of fish.

INT. FALSE CREEK OCEAN (2022) — DAY

Sticklebacks swim in a very large school.

                    NARRATOR (V.O.)
          And some research indicates
          sticklebacks are susceptible to
          sea lice infestations.

THOUSANDS OF STICKLEBACKS — swim by. There
are so many of them they are like a swarm.

                    NARRATOR (V.O.)
               But most research concludes
               that sticklebacks actually prey
               on sea lice.

HUNDREDS OF STICKLEBACKS — hover around a
piling.

                    NARRATOR (V.O.)
               So it's possible that
               sticklebacks can get infected
               with VHS and ENV or some other
               pathogen through interactions
               with sea lice.

CIRCLING A NET — sticklebacks hover around
the herring egg-covered panels.

THE STICKLEBACKS — move in to eat herring
eggs on a wrapped piling.

                    NARRATOR (V.O.)
               But in False Creek, there is
               compelling evidence that the
               sticklebacks became infected
               from eating herring eggs.

DOZENS OF STICKLEBACKS — swim up to the
herring eggs on a panel.

LOOKING DOWN — on the panel. There's a
feeding frenzy, stickleback-style, going on.

Sticklebacks in False Creek eat herring eggs laid on creosote pilings at Fisherman's Wharf. All stages of a herring's life play a crucial role in the marine food web. Eggs, larvae, immature herring and herring adults are key prey items on the Pacific coast. (Photo: Juggernaut Pictures Inc.)

Sticklebacks swimming around Fisherman's Wharf have white mucus trailing from their vents after eating herring eggs. More research is needed, but this seems to indicate they contracted a viral infection from eating herring eggs. (Photo: Scott Renyard)

> NARRATOR (V.O.)
> During the springs of 2021 and
> 2022,

FROM THE SIDE — Sticklebacks look at the
fresh herring eggs.

> NARRATOR (V.O.)
> . . . sticklebacks ate herring
> eggs laid on the pilings and
> nets at Fisherman's Wharf.

LOOKING DOWN A PILING — The sticklebacks peck
away at the eggs.

ONE STICKLEBACK — pokes at an egg.

MONTAGE — of sticklebacks with white mucus
trailing from their vents.

> NARRATOR (V.O.)
> Three weeks later, many of them
> had white mucus trailing from
> their vents, which is one tell-
> tale sign of infection. And
> lots of them were infected.

TWO STICKLEBACKS — with huge white mucus
trails hanging from their vents.

INT. FISHERMAN'S WHARF OCEAN — DAY

It's undeniable that the two species spend
significant time in close proximity to each
other. The sticklebacks are busy eating eggs
while herring spawners are laying eggs on
the same piling below them.

> NARRATOR (V.O.)
> This likely means that herring
> can spread pathogens to other
> species of fish at all stages
> of its life cycle. And if
> herring are able to do this,
> this has dire consequences for
> the entire marine ecosystem.

THOUSANDS OF STICKLEBACKS — mingle in the
green water off Fisherman's Wharf.

> NARRATOR (V.O.)
> Until wild and farmed fish are
> completely separated,

PANNING A FARM — Its pens are reflected in
the still ocean water. A metaphor perhaps for
how the farm is operating against itself.

UNDERWATER — Herring swim past the farm.

BACK TO THE FARM — Farmed salmon swim near
the surface. Their fins are sticking out of
the water.

                    NARRATOR (V.O.)
          . . . the farms will continue
          to provide ideal conditions for
          pathogens to flourish,

CLOSER — through a net. A farmed salmon leaps
into the air.

                    NARRATOR (V.O.)
          . . . which will continue to
          spread diseases into wild
          Pacific herring populations and
          the species that rely on them.

BACK UNDERWATER — Herring and farmed salmon
swim together.

THOUSANDS OF HERRING — mingle outside a fish
farm net pen.

EXT. WEST COAST OCEAN — DAY

Commercial fishing boats are spread out at
the fishing grounds.

                    NARRATOR (V.O.)
          We've all heard the saying
          "death by a thousand cuts."

CLOSER — The fishermen pull in their
gillnets.

                    NARRATOR (V.O.)
          But maybe for Pacific herring,
          it's "death by five cuts."

EXT. LAMBERT CHANNEL (1940s) — DAY

A fishing fleet is getting ready for an
opening.

IN A SEINE NET — The herring look for an
escape route.

> NARRATOR (V.O.)
> The first cut, overharvesting
> during the Reduction Fishery
> years that may have eliminated
> some herring populations.

UNDERWATER — A seine net full of herring is
pulled tight.

THE HERRING — look like a ball of fish
wrapped tightly in the net.

> NARRATOR (V.O.)
> The second cut,
> overexploitation of many
> herring populations by revenue-
> focused fisheries management.

FISHERMEN — brail a basket half-full of
herring from the skiff to the hold of a
larger vessel.

EXT. SQUAMISH ESTUARY — DAY

The area is dominated by logging operations
and development.

TRAVELLING ALONG THE WATER — Logs fill the channel.

> NARRATOR (V.O.)
> The third cut, spawning habitat degradation from shoreline development.

ON A PILING — Thousands of herring eggs are dying.

OIL-ABSORBENT PADS — float inside an oil spill response boom.

> NARRATOR (V.O.)
> The fourth cut, egg damage from oil and other contaminants.

FROM ABOVE (DRONE) — A massive 10-pen farm fills the bay.

> NARRATOR (V.O.)
> And the fifth and deepest cut, the amplification of lethal diseases from open net fish farms.

THROUGH A FARM — predator net. Young farmed salmon leap out of the water.

UNDERWATER — Herring dance their way through the layers of fish farm nets.

RESUME — Matsen interview.

                    JONN MATSEN
          We're volunteers,

EXT. FISHERMAN'S WHARF "A" FLOAT — DAY

Matsen is checking a panel placed in the
water for herring to spawn on.

                    JONN MATSEN (O.S.)
          . . . we're getting tremendous
          results,

CLOSE ON MATSEN — He shakes the silt off
another panel.

                    JONN MATSEN
          . . . but we want 100 percent
          results.

FROM ABOVE — A fish farm crew in a boat is
harvesting farmed fish.

                    JONN MATSEN (O.S.)
          If you want salmon farms, get
          them out of the water.

LOOKING THROUGH A NET — The farmed fish are
mature, swimming slowly.

                    JONN MATSEN (O.S.)
          They're destroying our salmon,
          they're destroying our herring.

FARMED FISH — are being sucked up into a
semi-transparent flex pipe.

                    JONN MATSEN (O.S.)
          We want our fishery back.

INT. NEAR FISH FARM OCEAN — DAY

Thousands of herring are mingling around the
farm.

                    JONN MATSEN (O.S.)
               Every sports fisherman in
               Canada is aghast that the
               federal government is not . . .

ANOTHER ANGLE — Thousands of herring are
swimming near a fish farm.

                    JONN MATSEN (V.O.)
               . . . doing its mandate to
               protect and enhance wild fish.

EXT. MILTY OCEAN SHORELINE — DAY

A school of herring fade in and out of the
milt right along the shoreline.

                    DOUG HAY (O.S.)
               Herring are one of the most
               abundant and probably important
               species that keep ecosystems
               going here.

A HERON — grabs a herring in its bill and
eats it.

HERRING — swim through a milty ocean.

                    DOUG HAY (O.S.)
          In fact, in terms of total
          abundance,

RESUME — Hay interview.

                    DOUG HAY
          . . . in terms of biomass,
          they're a much more important
          component in the environment
          than things like salmon.

INT. OCEAN — DAY

Beautiful, healthy herring cruise by through
clouds of milt.

                    DOUG HAY (O.S.)
          I think that forage fish are
          receiving more attention . . .

BACK TO HERRING — swimming close to camera in
spectacular fashion.

                    DOUG HAY (O.S.)
          . . . than they did in the past,
          but probably not enough and
          probably not fast enough.

EXT. FISHERMAN'S WHARF — DAY

Jonn Matsen is checking another net, giving
it a shake to clean off some silt.

ON MATSEN'S HANDS — He cleans another net.

               NARRATOR (V.O.)
     Since the streamkeepers have
     been monitoring herring in
     False Creek,

UNDERWATER — A large school of herring are
laying eggs on the net.

               NARRATOR (V.O.)
     . . . they've recorded up to
     five distinct spawning events
     per spring.

ANOTHER ANGLE — The spawning activity is
spectacular.

               NARRATOR (V.O.)
     Each one about a month apart.

THE HERRING — are spooked by something. They
stop spawning and disappear.

THE EGGS — remain on the mesh and appear to
be very much alive.

               NARRATOR (V.O.)
     These different spawn timings
     suggest that distinct herring
     populations still exist in
     small numbers,

FALSE CREEK HERRING — cruise by slowly.

>   NARRATOR (V.O.)
>   . . . even in the highly
>   urbanized Burrard Inlet system.

BACK TO — the herring spawning on a panel.

>   NARRATOR (V.O.)
>   This finding gives us hope
>   that some smaller herring
>   populations have survived and
>   still exist.

UP CLOSE — The herring have fully covered the
panel with eggs.

BACK TO MATSEN — He pulls up another panel
and gives it a shake.

>   NARRATOR (V.O.)
>   The streamkeeper work in False
>   Creek and . . .

EXT. SQUAMISH TERMINALS — DAY

Three streamkeepers pull a float line across
the Squamish Terminals east dock.

>   NARRATOR (V.O.)
>   . . . Howe Sound may seem
>   insignificant in the grand
>   scheme of things.

MATSEN — makes his way along the shoreline of
the Mamquam Blind Channel. He is looking for
herring eggs.

                    NARRATOR (V.O.)
          But maybe their targeted
          approach is needed to rebuild
          and bring back local herring
          populations, one inlet at a
          time.

TWO UNDERWATER SHOTS — of a thick herring
school moving slowly through the water.

ON A BARNACLED PILING — Herring are spawning
on it.

                    NARRATOR (V.O.)
          The streamkeepers have been
          trying to use the herring's
          natural behaviour to help
          them survive in compromised
          environments.

HERRING — swim straight up a piling and turn
at the surface.

AFTER THE SPAWN — a wrapped piling is covered
in eggs.

                    NARRATOR (V.O.)
          Their struggles to do so speak
          volumes about how hard it is
          for us to replicate nature.

UNDERWATER BANK — Herring are spawning on
marine plants.

MORE HERRING — are swimming through a cloud
of milt.

SEA LIONS — swim swiftly across a clean
ocean.

>   NARRATOR (V.O.)
>   Humans, like all animals, have
>   a deep-rooted instinct to
>   accumulate more than we need
>   in order to survive periods of
>   scarcity.

EXT. OCEAN — DAY

Sea lions frolic in the ocean. They are
searching for and feeding on herring.

A SEINER — bobs in the ocean. Hundreds of
gulls and bald eagles circle around it.

>   NARRATOR (V.O.)
>   But this instinct, when out of
>   control, is pushing nature to
>   the brink of collapse.

INT. ROCKWEED OCEAN — DAY

Beautiful rockweed sways back and forth in
the ocean.

WIDER — Herring swim in front of a beautiful
backdrop.

The first cut affecting herring populations was the overharvesting during the Reduction Fishery Era (1937–1967). The herring fishery was shut down by the Canadian Government, but some distinct herring populations may have been extirpated. (Photo: Fisherman Publishing Society)

The second cut affecting herring populations was overly aggressive harvesting of smaller resident herring populations. (Photo: Fisherman's Publishing Society)

The third cut affecting herring populations is alterations to the foreshore that eliminated safe spawning surfaces on which herring could lay their eggs. It now seems that herring eggs do best on marine or foreshore plants and their removal is likely a key factor in the decline of local populations that exist near degraded environments. (Photo: Ed Thwaites)

The fourth cut is exposure of herring eggs to oil and oily contaminants. This factor increases when herring eggs are laid on hard structures like pilings, which exposes them to oil floating on the surface of the water. (Photo: Scott Renyard)

The fifth and most serious cut was exposure to open net pen fish farms. The fish farms pose several hazards to herring, including abnormally high sea lice infestations, viral and bacterial pathogens, and the risk of becoming prey of the farmed fish when they swim into the pens looking for food. (Photo: Tavish Campbell)

This photo is an example of healthy eggs on bladder wrack kelp (*Fucus vesiculosus*). Shorelines with healthy plant life might be the most important factor for herring egg survival. (Photo: Scott Renyard)

                    WOODY MORRISON (O.S.)
          "If you go down into their
          world, you have to pass through
          that veil. And then you can see
          'em."

NOW — the herring cruise over a bed of
rockweed.

SOME HERRING — dart upward through a milty
ocean.

                    WOODY MORRISON (O.S.)
          "That's our relatives, that's
          Enong, that's the herring
          people." And he asks, "Who's
          the herring people?"

BACK TO ROCKWEED — loaded with eggs. It's
magical.

                    WOODY MORRISON (V.O.)
          "Well, they're the ones who put
          food out there for everybody."
          And he says, "How do they do
          that?"

HERRING — cruise through a *Sargassum muticum*
(also known as Japanese wireweed) patch.

A FILAMENTOUS ALGAE — known as feather boa
kelp, or *Egregia menziesii*, loaded with
herring eggs is drawn to the surface.

                    WOODY MORRISON (V.O.)
               "Where do they put it?"

THE KELP — is pulled upward.

                    WOODY MORRISON (V.O.)
               "It's in the water. When they
               offer the gift of food, we
               don't take it for granted, we
               accept it."

ABOVE WATER — Each string of kelp is covered
in eggs. A hand rearranges the eggs.

HERRING — are busy spawning on another patch
of japweed.

                    WOODY MORRISON (V.O.)
               "And when you're eating food,
               so long as you're enjoying it,
               you're treating it with the
               esteem required."

A LONE GILLNETTER — cruises across a purple
and pink sunset.

                    WOODY MORRISON (V.O.)
               "But if you continue past the
               point of enjoyment, now you're
               treating it disrespectfully."

EXT. EXPANSIVE OCEAN — DAY

Just a few fishing boats dot the landscape
far out in the ocean.

                                        FADE OUT:

EXT. SHORELINE — DUSK

Travelling over a shoreline that would
attract herring spawners.

**PLATE 1: The Squamish Streamkeepers
continue to support and monitor the herring
populations in Howe Sound and False Creek.**

**PLATE 2: Pacific herring populations continue
to struggle on British Columbia's coast,
requiring even stricter limits on commercial
harvesting.**

The tail credits begin.

TAIL CREDITS . . .

DON HERRING (O.S.)
Ships appear at night, ships
appear with lights.

Nets are cast, there's no
escape.

Swept up by my all, taken for
the pay.

It all happens without shame.

When daylight appears, I am
nowhere near.

Gone, only memory in my way.

I am the herring but
something's gone wrong.

My days are numbered, I've not
got long.

Listen to my call, listen to my
song.

I am the herring, help me 'fore
I'm gone.

You may say you know, you may
say it's fine.

But there are things you don't
see.
                    (MORE)

DON HERRING (O.S.) (CONT'D)
I am but a link, in a chain
that spans all time.

I wish you'd listen to my plea.

Stop this wanton lust.

The boom has gone to bust.

Your life is tied to me.

I am the herring, but
something's gone wrong.

My days are numbered, I've not
got long.

Listen to my call, listen to my
song.

I am the herring, help me 'fore
I'm gone.

I am the herring, help me 'fore
I'm gone.

I am the herring, help me 'fore
I'm gone.

PLATE 3: In memory of Woodrow "Woody" Friedlander Morrison, Jr.

December 22, 1941-January 28, 2021

"A respected Haida elder and wonderful storyteller."

PLATE 4: In memory of Thorkild "Thor" Dissing Froslev

March 15, 1933-September 12, 2022

"We will all miss his unrelenting support of artists and the environment."

THE END

Acknowledgements

A great big thank-you to all of the Squamish Streamkeepers—Jonn Matsen, Jack Cooley, Cal Hartnell, Patrick MacNamara, Brad Ray, Douglas Swanston, Hugh Kerr, Lyle Wood, Keith Pellettier, Don Lawrence, Ted Domachowski, Roy Sakata, Don Robson, Ana Santos, Eric Andersen, Pam Tattersfield, Alison Wald, Spencer Fischen, Noel Murphy, Jane Smalley, Zoe Blue, Brandon Hartnell, and Keith Wright—not only for their cooperation during many days of filming, but also for their perseverance and dedication to the herring of Squamish and False Creek. If they had quit or wavered at all in their resolve, my film *The Herring People* would not have been possible.

I am very grateful to all of my interview subjects, namely Woody Morrison, Dr. Doug Hay, Dr. Jeff Marliave, Percy Redford, Grant Scott, Calvin Siider, Dr. Thomas E. Reimchen, David Ellis, Dr. Jonn Matsen, John Buchanan, Eric Andersen, Edith Tobe, Douglas Swanston, Fred Felleman, Kurt Stick, Dr. Richard Steiner, Dr. Gary Marty, and Dr. Gideon Mordecai. I thank all of them for sharing their time and knowledge about Pacific herring.

I would like to thank the staff and management of the Squamish Terminals for their support for the Squamish Streamkeeper projects and my film making. Access to the Terminals was essential to the making of the film. It's not often that large corporations cooperate to such an extent on projects that could add costs or attract criticism. Their participation helped shed light on the behaviour of Howe Sound herring and helped the company engineer a safer terminal for herring after the 2015 fire. This was a great example of how corporations can lessen their impacts on the environment.

I would also like to thank the staff and management of Fisherman's Wharf, located in False Creek. I had free and open access at any time to film around the floats and pilings. I would not have been able to capture the different parts of the herring life cycle and the species that were preying on them without year-round access.

A special thanks to Dr. Jonn Matsen. He had the original idea to investigate the Squamish herring and to see if they could be brought back to the levels he remembered from his childhood. After the discovery that herring eggs were dying on creosote pilings, he sourced the wrapping materials and designed the float lines and always seemed to take setbacks as a positive step toward finding the best solution. In addition, John Buchanan's surveys added tremendously to the streamkeepers' knowledge about the spawning range in Howe Sound. These surveys led to the discovery that hard surfaces like rocks might contribute to herring egg mortality and pushed the streamkeepers to find ways to keep herring eggs submerged and shielded from exposure to air.

A great big thanks, as always, to Jan Westendorp and Lesley Cameron, who make my book projects possible: Jan for her amazing designs and knowledge of the publishing process. I would be lost in a minute without her expertise. And Lesley for her tremendous edits and attention to detail. Nothing gets past her watchful eye. I'm so very grateful to both of you. I would also like to thank Caid Dow, who brought my vision of the film poster to life. It took some doing to show the world what the little herring people look like and how amazing their eyes are.

A special thanks to the City of Vancouver Archives, the Squamish Library Archives, the Alaska State Archives, and the University of British Columbia Special Collections and their amazing librarians and staff for their help in locating historical photographs and footage. And last but not least, a special mention to Eric Andersen for his help in locating and expertise regarding historical images of the Squamish Estuary and Mamquam Blind Channel.

Bibliography

Entries in bold are referrred to directly in the screenplay.

A look back at the Squamish harbor dredging. (2021, February 2). *Squamish Reporter.*

A very scenic exit for B.C.'s exports. (1974, July–August). *Supply Post, 2*(9).

Beacham, T.D., Schweigert, J.F., MacConnachie, C., Le, K.D., & Flostrand, L. (2008). Use of microsatellites to determine population structure and migration of Pacific herring in British Columbia and adjacent regions. *Transactions of the American Fisheries Society, 137*: 1795–1811.

Bell, L.M. (1975). *Factors influencing the sedimentary environments of the Squamish River delta in southwestern British Columbia* [Master of Applied Science thesis, University of British Columbia]. UBC Theses and Dissertations.

Bell-Irving, R. (1989, February 1). Fishery permitted without any study [Letter to the editor]. *Vancouver Sun.*

Bennett, K. (2003, May). *Haegele eelgrass metadata report: Source metadata and digital data specifications.* Geostreams Consulting Report.

Benson, A.J., Cox, S.P., & Cleary, J.S. (2015, July). Evaluating the conservation risks of aggregate harvest management in a spatially-structured herring fishery. *Fisheries Research, 167*: 101–113.

Bodtker, K. (2017). *Ocean Watch: Howe Sound edition. Coastal Ocean Research Institute report.* Coastal Ocean Research Institute.

Boldt, J.L., Therriault, T.W., Hay, D.E., Schweigert, J., & Thompson, M. (n.d.). Nearshore fish community dynamics in the Strait of Georgia: Information from juvenile herring surveys [PowerPoint presentation]. Fisheries and Oceans Canada, Science Branch, Pacific Biological Station, Nanaimo, BC, Canada.

Boldt, J.L., Thompson, M., Rooper, C.N., Hay, D.E., Schweigert, J.F., Quinn, T.J., Cleary, J.S., & Neville, C.M. (2018). Bottom-up and top-down control of small pelagic forage fish: Factors affecting age-0 herring in the Strait of Georgia, British Columbia. *Marine Ecology Progress Series, 617–618*: 53–66.

Brodeur, L.P. (1906, February). Thirty-eighth annual report of the Department of Marine and Fisheries. *Sessional Paper No. 22.* Department of Marine and Fisheries.

Busch, P. (1992, September 15). BCR port could be deep sixed. *Squamish Chief*, 2(37).

Carrothers, W.A. (1941). *The British Columbia fisheries.* The University of Toronto Press.

Castagna, M. (1973). Shipworms and other marine borers. *Marine Fisheries Review, 35*(8): 7–12.

Christensen, V. (2010). Behaviour of sandeels feeding on herring larvae. *The Open Fish Science Journal, 3*: 164–168.

Cleary, J., & Grimnell, M. (2020, July 30). *Pacific herring preliminary data summary for Strait of Georgia 2020.* Pacific Biological Station, Fisheries and Oceans Canada.

Cleaver, F.C., & Franett, D.M. (1946). *The predation by sea birds upon the eggs of the Pacific herring* (Clupea pallasii) *at Holmes Harbor during 1945.* Biological Report No. 46B. State of Washington Department of Fisheries.

Commission of Inquiry into the Decline of Sockeye Salmon in the Fraser River. (2011, August 31). Public hearings transcripts. Government of Canada (p. 56).

Davis, J.C. (1989, June 7). *Nestucca oil spill.* Department of Fisheries and Oceans—Report on Spill Response.

Department of Fisheries. (1937). *Seventh Annual Report of the Department of Fisheries: For the year 1936–37.* **Dominion of Canada.**

Director General. (1991, June 11). Memorandum RE: Federal-provincial policy for the importation of Atlantic salmon into British Columbia. Fisheries and Oceans Canada. Government of Canada.

Dolbeth, M., Crespo, D., Leston, S., & Solan, M. (2019). Realistic scenarios of environmental disturbance lead to functionally important changes in benthic species-environment interactions. *Marine Environmental Research,* (150): 1–7.

Dredging to deepen small boat harbour: Work goes on 24 hours a day. (1959, January 15). *Squamish Times.*

Dunn, F. (2018, January 4). Freedom of Information documents: RE: ISA Nova Scotia Fisheries and Aquaculture Office of the Deputy Minister.

Emmett, R.L., Stone, S.L., Hinton, S.A., & Monaco, M.E. (1991). *Distribution and abundance of fishes and invertebrates in west coast estuaries. Volume II: Species life history summaries.* ELMR Report Number 8. NOAA/NOS Strategic Environmental Assessments Division, Rockville, MD.

Exxon Valdez Oil Spill Trustee Council. (2008, January 10). *Prince William Sound herring restoration plan* [Draft]. Alaska Department of Fish and Game.

First ship expected here in October: Squamish Port called the most modern in western Canada. (1972, August 16). *Squamish Citizen.*

Fisheries and Marine Service. (1975, January–February). Herring spawn on kelp—1975. *Sounder Magazine.*

Fisheries and Oceans Canada. (2002). *Integrated fisheries management plan: Roe herring February 10–April 30, 2002.* Pacific Region. Government of Canada.

Fisheries and Oceans Canada. (2021). British Columbia herring spawn locations. *Geographical Bulletin, Pacific Region.* Government of Canada.

Fisheries and Oceans Canada. (n.d.). *Tenderfoot Creek Hatchery Chinook Conservation Program.* Government of Canada.

Flinkman, J., Aro, E., Vuorinen, I., & Viitasalo, M. (1998). Changes in northern Baltic zooplankton and herring nutrition from 1980s to 1990s: Top-down and bottom-up processes at work. *Marine Ecology Progress Series, 165*: 127–136.

Foerster, R.E. (1941). *Annual report of the Pacific Biological Station for 1941.* **Fisheries Research Board of Canada. Nanaimo, BC.**

Fort, C., Daniel, K., & Thompson, M. (2009, January). *Herring spawn survey manual.* Fisheries and Oceans Canada.

Fox, C.H., Paquet, P.C., & Reimchen, T.E. (2018). Pacific herring spawn events influence nearshore subtidal and intertidal species. *Marine Ecology Progress Series, 595*: 157–169.

Garver, K.A., & Hawley, L.M. (2021, January). *Characterization of viral haemorrhagic septicaemia virus (VHSv) to inform pathogen transfer risk assessments in British Columbia.* DFO Canadian Science Advisory Secretariat Research Document. 2020/064.

Garver, K.A., Traxler, G.S., Hawley, L.M., Richard, J., Ross, J.P., & Lovy, J. (2013). Molecular epidemiology of viral haemorrhagic septicaemia (VHSv) in British Columbia, Canada, reveals transmission from wild to farmed fish. *Diseases of Aquatic Organisms, 104*: 93–104.

Gorge Waterway Initiative. (2007, May). Fish of the waterway [GWI info sheet].

Government of Canada. (1948). *Progress reports of the Pacific Coast Station index.*

Government of Canada. (2023). *Pacific herring 2022–2023: Integrated fisheries management plan summary.*

Gozlan, R.E., Marshall, W.L., Lilje, O., Jessop, C.N., Gleason,
F.H., & Andreou, D. (2014, February 19). Current ecological
understanding of fungal-like pathogens of fish: What lies
beneath? *Frontiers in Microbiology, 5*(62): 1–16.

Gustafson, R.G., Drake, J., Ford, M.J., Myers, J., Holmes, E.E., & Waples,
R.S. (2006). Multidisciplinary examination of Pacific herring
(*Clupea pallasii*) population discreteness: The Cherry Point
population and the usa's *Endangered Species Act*. U.S. Department
of Commerce, noaa Technical Memo. nmfs-nwfsc-76.

Haist, V., Schweigert, J.F., & Stocker, M. (1986, June). *Stock assessments
for British Columbia herring in 1985 and forecasts of the potential catch
in 1986.* Canadian Manuscript Report of Fisheries and Aquatic
Sciences. No. 1889.

Hansen, B.H, Salaberria, I., Read, K.E., Wold, P.A., Hammer, K.M., Olsen,
A.J., Altin, D., Overjordet, I.B., Nordtug, T., Bardal, T., & Kjorsvik,
E. (2019, July 3). Developmental effects in fish embryos exposed
to oil dispersions: The impact of crude oil micro-droplets. *Marine
Environmental Research, 150*: 1–12.

Hay, D.E. (1986). Effects of delayed spawning on viability of eggs and
larvae of Pacific herring. *Transactions of the American Fisheries
Society, 115*(1): 155–161.

Hay, D.E., & Kronlund, A.R. (1987). Factors affecting the distribution,
abundance, and measurement of Pacific herring (*Clupea harengus
pallasii*) spawn. *Canadian Journal of Fisheries and Aquatic Sciences,
44*(6): 1181–1194.

Hay, D.E., & McCarter, P.B. (2013, revised 2019). *Herring spawn areas
of British Columbia: A review, geographic analysis and classification.*
Volumes 1–6. Canadian Manuscript Report of Fisheries and
Aquatic Sciences.

Hay, D.E., McCarter, P.B., Kronlund, R., & Roy, C. (1989). *Spawning areas of British Columbia herring: A review, geographical analysis and classification.* Volume V. Canadian Manuscript Report of Fisheries and Aquatic Sciences, No. 2019.

Hershberger, P.K., Elder, N.E., Wittouck, J., Stick, K., & Kocan, R.M. (2011). Abnormalities in larvae from the once-largest Pacific herring population in Washington State result primarily from factors independent of spawning location. *Transactions of the American Fisheries Society, 134*(2): 326–337.

Hershberger, P.K., Garver, K.A., & Winton, J.R. (2016). Principles underlying the epizootiology of viral hemorrhagic septicemia in Pacific herring and other fishes throughout the North Pacific Ocean. *Canadian Journal of Fisheries and Aquatic Sciences, 73*(5): 853–859.

Hourston, A.S. (1978, June). *The decline and recovery of Canada's Pacific herring stocks.* Fisheries and Marine Service Technical Report 784.

Hume, M. (2012, September 6). Howe Sound: A herring revival spawned from the depths of darkness. *Globe and Mail.*

Hume, S. (2009, April 2). A silvery sliver of good environmental news — on False Creek. *Vancouver Sun.*

Keddie, G. (n.d.). A Lekwungen herring fishing site in Esquimalt Harbour [Undated collection of notes]. Royal BC Museum.

Kiorboe, T., Munk, P., Richardson, K., Christensen, V., & Paulsen, H. (1988). Plankton dynamics and larval herring growth, drift and survival in a frontal area. *Marine Ecology—Progress Series, 44*: 205–219.

Kocan, R.M. (2018). Transmission models for the fish pathogen Ichthyophonus: Synthesis of field observations and empirical studies. *Canadian Journal of Fisheries and Aquatic Sciences, 76*(4): 636–642.

Kocan, R.M., Hershberger, P., Mehl, T., Elder, N., Bradley, M., Wildermuth, D., & Stick, K. (1999, January 7). Pathogenicity of Ichthyophonus hoferi for laboratory-reared Pacific herring *Clupea pallasi* and its early appearance in wild Puget Sound herring. *Diseases of Aquatic Organisms, 35*(1):23–29.

Larsen, E.M. (2004, October). *Nestucca oil spill: Revised restoration plan.* U.S. Fish and Wildlife Service, Western Washington Fish and Wildlife Office, Lacey, Washington.

Lee, B. (2009, April). A glimmer of hope: The Pacific herring and its connection to Pender Harbour. *Harbour Spiel* (20):12–16.

Lindquist, A., & Sandell, T. (2018). *Cherry Point herring update, 2018.* Washington Department of Fish and Wildlife.

Love, B., Villalbos, C., & Olson, M.B. (2018). Interactive effects of ocean acidification and ocean warming on Pacific herring *(Clupea pallasii)* early life stages, *Salish Sea Ecosystem Conference proceedings.*

Lowry, D. (Ed.). (2019, April 23–24). *Washington Department of Fish and Wildlife Contribution to the 2019 Meeting of the Technical Sub-committee (TSC) of the Canada-U.S. Groundfish Committee: Reporting for the period from May 2018–April 2019.* **Washington Department of Fish and Wildlife, Olympia, Washington.**

MacLeod, J.R. (1972, July). *The herring fishery: Potential for expansion. Report of the Herring Task Force.* **Fisheries and Oceans Canada, Pacific Region.**

Marliave, J.B. (1975). Behaviour transformation from the planktonic larval stage of some marine fishes reared in the laboratory [Doctoral dissertation, Department of Zoology, University of British Columbia]. UBC Theses and Disserations.

Martell, J.D., Duhalme, J., & Parsons, G.J. (Eds.). (2013). *Canadian aquaculture review R&D review 2013.* Aquaculture Association of Canada Special Publication.

Martin, J.R. (2013). Turnover time and annual migration of Clupea harengus Atlantic herring on the German Bank of Scot's Bay spawning grounds: A mark and recapture study. [Master of Science thesis, the University of New Brunswick].

Marty, G.D., Freiberg, E.F., Meyers, T.R., Wilcock, J., Farver, T.B., & Hinton, D.E. (1998). Viral hemorrhagic septicemia virus, *Ichthyophonus hoferi*, and other causes of morbidity in Pacific herring *Clupea pallasi* spawning in Prince William Sound, Alaska, USA. *Diseases of Aquatic Organisms*, 32(1):15–40.

Marty, G.D., Hulson, P.F., Miller, S.E., Quinn, T.J., Moffitt, S.D., & Merizon, R.A. (2010). Failure of population recovery in relation to disease in Pacific herring. *Diseases of Aquatic Organisms*, 90(1):1–14.

McHugh, J.L. (1942). Variation of vertebral centra in young Pacific herring (*Clupea pallasii*). *Journal of the Fisheries Research Board of Canada*, 5(4): 347–360.

McKechnie, I., Lepofsky, D., Moss, M.L., Butler, V.L., Orchard, T.J., Coupland, G., Foster, F., Caldwell, M., & Lertzman, K. (2014, March 4). Archaeological data provide alternative hypotheses on Pacific herring (*Clupea pallasii*) distribution, abundance, and variability. *Proceedings of the National Academy of Sciences of the United States of America*, 111(9): E807–E816.

Meyers, T., Burton, T., Bentz, C., Ferguson, J., Stewart, D., & Starkey, N. (2019, July). *Diseases of wild and cultured fishes in Alaska: Alaska Department of Fish and Game fish pathology report*. Alaska Department of Fish and Game.

Mordecai, G.J., Cicco, E.D., Gunther, O.P., Schulze, A.D., Kaukinen, K.H., Li, S., Tabata, A., Ming, T.J., Ferguson, H.W., Suttle, C.A., & Miller, K.M. (2020). Discovery and surveillance of viruses from salmon in British Columbia using viral immune-response bio-markers, metatranscriptomics, and high-throughput RT-PCR. *Virus Evolution*, 7(1): 1–14.

Mordecai, G.J., Miller, K.M., Cicco, E.D., Schulze, A.D., Kaukinen, K.H., Ming, T.J., Li, S., Tabata, A., Teffer, A., Patterson, D.A., Ferguson, H.W., & Suttle, C.A. (2019, September). Endangered wild salmon infected by newly discovered viruses. *eLife*, 1–18.

Munk, P., Kiorboe, T., & Christensen, V. (1989). Vertical migrations of herring, *Clupea harengus*, larvae in relation to light and prey distribution. *Environmental Biology of Fishes, 26*(2): 87–96.

Naumenko, N.I. (2002, August). *Temporal variations in size-at-age of the western Bering Sea herring. Pices-Globec International Program on Climate Change and Carrying Capacity.* PICES Scientific Report No. 20. North Pacific Marine Science Organization.

Nautilus Environmental. (2008, April). *Environmental Monitoring Program for the Nanaimo River Estuary.* Draft report for the City of Nanaimo and Nanaimo Estuary Management Committee, Nanaimo, BC.

Nylund, A., Brattespe, J., Plarre, H., Kambestad, M., & Karlsen, M. (2019). Wild and farmed salmon (*Salmo salar*) as reservoirs for infectious salmon anemia virus, and the importance for horizontal and vertical transmission. *PLoS ONE, 14*(4): e0215478.

Nylund, A., Hovland, T., Hodneland, K., Nilsen, F., & Løvik, P. (1994). Mechanisms for transmission of infectious salmon anaemia (ISA). *Diseases of Aquatic Organisms, 19*: 95–100.

Nylund, A., Wallace, C., & Hovland, T. (1993). The possible role of Lepeophtheirus salmonis in the transmission of infectious salmon anaemia. In G. Boxhall (Ed.), *Pathogens of wild and farmed fish: Sea lice* (pp. 236–240). Ellis Horwood Ltd.

Pagowski, V.A., Mordecai, G.J., Miller, K.M., Schulze, A.D., Kaukinen, K.H., Ming, T.J., Li, S., Teffer, A.K., Tabata, A., & Suttle, C.A. (2019). Distribution and phylogeny of erythrocytic necrosis virus (ENV) in salmon suggests marine origin. *Viruses, 11*(358): 1–16.

Parsons, G.J., Burgetz, I.J., Weber, L., Garver, K.A., Jones, S.R.M., Johnson, S., Hawley, L.M., Davis, B., Aubry, P., Wade, J., & Mimeault, C. (2021). *Assessment of the risk to Fraser River sockeye salmon due to viral haemorrhagic septicaemia virus IVa (VHSV-IVa) transfer from Atlantic salmon farms in the Discovery Islands area, British Columbia.* Department of Fisheries and Oceans, Canadian Science Advisory Secretariat Document. 2020/065.

Pauly, D., & Christensen, V. (1995). Primary production required to sustain global fisheries. *Nature, 374*: 255–257.

Pearson, W.H., Deriso, R.B., Elston, R.A., Hook, S.E., Parker, K.R., & Anderson, J.W. (2011). Hypotheses concerning the decline and poor recovery of Pacific herring in Prince William Sound, Alaska. *Reviews in Fish Biology and Fisheries, 22*(2):95–135.

Penn, Briony. (2020, April 28). Courting collapse. *Focus on Victoria.*

Petrou, E.L., Fuentes-Pardo, A.P., Rogers, L.A., Orobko, M., Tarpey, C., Jiménez-Hidalgo, I., Moss, M.L., Yang, D., Pitcher, T.J., Sandell, T., Lowry, D., Ruzzante, D.E., & Hauser, L. (2021). Functional genetic diversity in an exploited marine species and its relevance to fisheries management. *Proceedings of the Royal Publishing Society, 288*: 1–9.

Purcell, J.E., Grosse, D., & Grover, J.J. (1990). Mass abundances of abnormal Pacific herring larvae at a spawning ground in British Columbia. *Transactions of the American Fisheries Society, 119*(3): 463–469.

Ritter, J. (1999, June). Interaction between wild and farmed Atlantic salmon in the Maritime provinces. *Proceedings of the Diadromous Subcommittee Regional Advisory Process.* Department of Fisheries and Oceans, Science Branch, Maritime Region, Moncton, NB.

Rooper, C.N., Haldorson, L.J., & Quinn, T.J. (1998). An egg-loss correction for estimating spawning biomass of Pacific herring in Prince William Sound, Alaska. *Alaska Fishery Research Bulletin, 5*(2): 1–9

Salish Sea Pacific Herring Assessment and Management Strategy
Team. (2018). *Assessment and management of Pacific herring in the
Salish Sea: Conserving and recovering a culturally significant and
ecologically critical component of the food web.* The SeaDoc Society,
Orcas Island, WA.

Schweigert, J.F., & Linekin, M. (1990). *The Georgia and Johnstone
Straits herring bait fishery in 1986: Results of a questionnaire survey.*
Canadian Technical Report of Fisheries and Aquatic Sciences
No. 1721.

Schweigert, J.F., Boldt, J., Flostrand, L., & Cleary, J.S. (2010). A review of
factors limiting recovery of herring stocks in Canada. *ICES Journal
of Marine Science, 67*(9): 1903–1913.

Shaw, P.W., Turan, C., Wright, J.M., O'Connell, M., & Carvalho, G.R.
(1999). Microsatellite DNA analysis of population structure in
Atlantic herring (*Clupea harengus*), with direct comparison to
allozyme and mtDNA RFLP analyses. *Heredity, 83*: 490–499.

Small, M.P., Loxterman, J.L., Frye, A.E., Von Bargen, J.F., Bowman, C., &
Young, S.F. (2005, August). Temporal and spatial genetic structure
among some Pacific herring populations in Puget Sound and the
southern Strait of Georgia. *Transactions of the American Fisheries
Society, 134*(5): 1329–1341.

Speller, C.F., Hauser, L., Lepofsky, D., Moore, J., Rodrigues, A.T., Moss,
M.L., McKechnie, I., & Yang, D. (2012). High potential for using
DNA from ancient herring bones to inform modern fisheries
management and conservation. *PLoS ONE, 7*(11): e51122.

Squamish River Watershed Society. (2008, May 2). Squamish Estuary/
Mamquam Blind Channel Restorations 2007/2008 Final Report.

Squamish! Deep sea port: 45,000 ton forest products ships will load
here. (1967, December 6). *Squamish Times.*

Stick, K.C., & Lindquist, A. (2009, November). *2008 Washington State
herring stock status report.* Washington Department of Fish and
Wildlife. Stock Status Report No. FPA 09–05.

Stiffler, L. (2013, March 29). The real story of Puget Sound's disappearing herring. Sightline Institute.

Taylor, F.H.C. (1963). *The stock-recruitment relationship in British Columbia herring populations.* Technical Report, Fisheries Research Board of Canada, Biological Station, Nanaimo, BC.

Taylor, F.H.C. (1964). *Life history and present status of British Columbia herring stocks.* **Bulletin No. 143. Fisheries Research Board of Canada.**

Taylor, F.H.C. (1973). *Data record. Detailed tagging and tag recovery records of herring, 1957 to 1967.* **Fisheries Research Board of Canada, Manuscript Report Series No. 1262.**

Tester, A.L. (1949). Populations of herring along the west coast of Vancouver Island on the basis of mean vertebral number, with a critique of the method. *Journal of the Fisheries Research Board of Canada,* 7(7): 403–427.

Thompson, J.A.J., & McComas, F.T. (1973). *Distribution of mercury in the sediments and waters of Howe Sound, British Columbia.* Fisheries Research Board of Canada. Technical Report No. 396.

Traxler, G.S., Kieser, D., & Evelyn, T.P.T. (1995). Isolation of North American strain of VHS virus from farmed Atlantic salmon. *Aquaculture Update, 72.*

Traxler, G.S., Kieser, D., & Richard, J. (1999). Mass mortality of pilchard and herring associated with viral hemorrhagic septicemia virus in British Columbia, Canada. *American Fisheries Society/Fish Health Section Newsletter,* 27(4): 4–5.

Traxler, G.S., Roome, J.R., Lauda, K.A., & LaPatra, S. (1997). The appearance of infectious hematopoietic necrosisvirus (IHNV) and neutralizing antibodies in sockeye salmon (*Oncorhynchus nerka*) during their migration and maturation period. *Disease of Aquatic Organisms,* 28(1): 31–38

Van Walraven, A., & Buchanan, J. (2015, November 18). Pacific herring spawn surveys: Howe Sound North 2010–2015. Concerned Citizens Bowen, Bowen Island, BC.

Vines, C.A., Robbins, T., Griffin, F.J., & Cherr, G.N. (2000). The effects of diffusible creosote-derived compounds on development in Pacific herring (*Clupea pallasii*). *Aquatic Toxicology*, 51(2): 225–239.

Weldwood fined for fish kill. (1974, March 14). *Squamish Times.*

Winton, J., Kurath, G., & Batts, W. (2007, July). *Detection of viral hemorrhagic septicemia virus.* U.S. Geological Survey. Western Fisheries Research Center Report.

www.ingramcontent.com/pod-product-compliance
Lightning Source LLC
Chambersburg PA
CBHW052107030426
42335CB00025B/2880